MEMOIRS
OF THE
AMERICAN MATHEMATICAL SOCIETY

Number 126

THE EXISTENCE OF VALUE IN DIFFERENTIAL GAMES

by

ROBERT J. ELLIOTT and NIGEL J. KALTON

Published by the
American Mathematical Society
Providence, Rhode Island
1972

AMS 1970 Subject Classifications 90D05, 90D25
Key Words and Phrases: differential games, strategies, values, minimax conditions, relaxed controls, saddle points

Library of Congress Cataloging in Publication Data

Elliott, Robert James, and Nigel J. Kalton
The existence of value in differential games.

(Memoirs of the American Mathematical Society, no. 126)
Bibliography: p. [illegible]
1. Differential games. 2. Control theory. I. Kalton, Nigel J., joint author. II. Title. III. Series: American Mathematical Society. Memoirs, no. 126.
QA3.A57 no. 126 [QA272] 510'.8s [519'.3] 72-4562
ISBN 0-8218-1826-8

Printed in the United States of America

THE EXISTENCE OF VALUE IN DIFFERENTIAL GAMES

Robert J. Elliott and Nigel J. Kalton*†

Mathematics Institute
University of Warwick
Coventry, CV4 7AL.
United Kingdom.

*Present address of the second author:

Department of Pure Mathematics
University College of Swansea
Swansea, SA2 8PP.
United Kingdom.

†During the period of this research both authors were supported by S.R.C. Research Fellowships.

Received by the Editors 11 November, 1971.

THE EXISTENCE OF VALUE IN DIFFERENTIAL GAMES

Robert J. Elliott and Nigel J. Kalton

ABSTRACT

In the manner described in the Introduction we show the existence of value for all two person, zero-sum differential games of prescribed duration. Using the concept of relaxed controls from control theory we relate the approaches to differential games of A. Friedman (J. Differential Equations 7 (1970), 69-91) and W. Fleming (J. of Maths. and Mechanics 13 (1964) 987-1008). We show that if the 'Isaacs' condition' (see §5 below) is satisfied then the game has a value in the sense of Friedman. Over the relaxed controls the Isaacs' condition is always satisfied and so the game always has a value in this setting. We do not need Friedman's hypothesis that the two sets of control variables appear separated in the dynamical equations and payoff. The introduction of probabilistic ideas into differential games by relaxed controls thus gives a value, as the introduction of mixed strategies by von Neumann does for two person zero-sum matrix games.

These results were announced in "Values in Differential Games", Bull. Amer. Math. Soc., Vol.8 No.3 (1972).

CONTENTS

ACKNOWLEDGEMENTS

We would like to take this opportunity to thank Professor L. Markus for many stimulating discussions on the subject of this memoir. We would also like to thank Mrs G. Maddocks for typing the text.

VALUE IN DIFFERENTIAL GAMES

INTRODUCTION

Two person, zero-sum game theory was developed to study competitive contests in a static situation. Optimal control theory, on the other hand, investigates one player optimization problems in dynamical situations, that is, in situations described by differential equations. The subject of this paper is two person, zero-sum differential games and it can be considered as control theory with two opposing controllers or players. We, therefore, consider a dynamical situation described by differential equations, and, at the end of some fixed time, say after one unit of time, a payoff is computed. The game is zero-sum, meaning one player is trying to maximize and the other to minimize the payoff.

Differential games were first studied by Isaacs in the 1950's, but Isaacs did not publish his results until his book [13] appeared in 1965. One of Isaacs' main contributions was to derive heuristically a differential equation that the value of the game should satisfy. Unfortunately this so-called Isaacs-Bellman equation is nonlinear and highly degenerate and we cannot say in general that it has a solution. Mention should also be made here of Berkovitz, who studied differential games using methods similar to those of the calculus of variations. See, for example, [1].

Throughout a differential game each player has a continuum of moves and a strategy for each player is, roughly speaking, a rule telling the player what to do next on the basis of what has happened in the game so far. Because time is continuous difficulties arise in making this notion precise; in a sequence of papers [4], [5], [6], [7], Fleming circumvented this problem and studied differential games by considering a sequence of games with discrete time, and also by approximating the differential equations by difference equations. Each approximating game has an upper value W_n^+, and a lower value W_n^- depending on whether the minimizing or maximizing player moves first at each step.

There are then two problems: first, do W_n^+ and W_n^- tend to limits W^+ and W^- as the time $\delta = 1/2^{-n}$ between steps approaches zero; this is called the 'convergence problem' by Fleming: second, are, in fact, these two limits equal: i.e.: does $W^+ = W^-$, so that the game has a 'value'?

Using discrete approximation methods Fleming [4] first gave positive answers to both the above questions when the opposing control variables are separated in both the differential equations and the payoff, (see §12 below).

Although it is not known whether the Isaacs-Bellman equation has solutions, if a small amount of 'noise' is introduced into the game, the upper (or lower) value functions will satisfy a non-linear parabolic equation and, in the situation under investigation, work of Friedman [8] tells us that such equations do have unique solutions. Therefore, for more general differential games Fleming [5], [6] introduced 'noise' into the game and so into the approximating discrete difference games. He was then able to show that the upper and lower values of the approximating games approached the solution of the corresponding non-linear parabolic equation, and was also able to prove the convergence of these values as the amount of noise decreased to zero. Thus again his convergence problem had a positive answer. However, the functions in Fleming's differential equations and payoff had to satisfy a constant Lipschitz condition.

Following earlier work by Varaiya and Lin [17], in a recent series of papers [10], [11], [12], Friedman has studied differential games directly, not approximating them by difference equations. Friedman, however, did find it necessary to approximate the idea of a strategy by upper and lower strategies varying at only a finite number of division points in the interval (see §3).

Again, depending on which player chooses his control function over the next interval first at each division point Friedman obtains upper and lower values V_n^+, V_n^-. It is immediate from Friedman's definitions that V_n^+, V_n^- are monotonic decreasing (respectively, increasing), and so they converge to limits V^+ and V^- (not necessarily the same as W^+ and W^-). However, in order to prove his game has 'value', that $V^+ = V^-$, Friedman also has to assume the restrictive hypotheses that the opposing control variables appear 'separated' in both the differential equations and payoff, (again see §12 below). Friedman, though, only requires that the functions occurring satisfy weaker, variable Lipschitz conditions, and his payoff is more general than Fleming.

Von Neumann [18] introduced the idea of a mixed strategy, a probability over the possible plays, in order to obtain a value for two person zero-sum matrix games. Relaxed controls are obtained by choosing a probability measure on the space of control values at each point in time. Young [21] introduced relaxed controls into the calculus of variations and Warga [20] and McShane [14] applied them in control theory. In an earlier paper with L. Markus [3] we have used relaxed controls to obtain saddle points for certain linear games.

In the paper below, following Roxin [16], we first define strategies and then give a re-formulation of Friedman's results. Using certain approximating games and relaxed controls, without any 'separation' hypotheses, we relate Friedman's upper and lower values to those of Fleming.

Thus, our principal results are that if the 'Isaacs' condition' is satisfied the game has a value in the sense of Friedman. Over the relaxed controls the 'Isaacs' condition' is always satisfied and so the game always has a value in this setting. These results are first proved under the 'constant Lipschitz' conditions of Fleming, but, by approximation arguments we are able to prove our results under the weaker hypotheses and for the more general payoff of Friedman. The paper ends with a discussion of saddle points for approximate strategies and the attainability of the value.

1. DEFINITION OF THE GAME G

We consider a differential game G played by two players J_1 and J_2 for the fixed time interval $I = [0,1]$. At each time $t \in I$, J_1 picks an element $y(t)$ from a fixed topological space Y, which is compact and metrizable, and J_2 picks $z(t)$ from a similar space Z, in such a way that the functions $t \to y(t)$ and $t \to z(t)$ are measurable (i.e. for continuous maps $\varphi: Y \to R$ and $\psi: Z \to R$, the functions $t \to \varphi(y(t))$ and $t \to \psi(z(t))$ are measurable in the ordinary Lebesgue sense). The *dynamics* of the game are given by the differential equation

$$\frac{dx}{dt} = f(t,x,y(t),z(t)). \tag{1}$$

Here $x \in R^m$ and

$$f: I \times R^m \times Y \times Z \to R^m$$

is a continuous function satisfying a Lipschitz condition

$$\|f(t,x_1,y,z) - f(t,x_2,y,z)\| \leq k(t)\|x_1 - x_2\| \tag{2}$$

whenever $x_1, x_2 \in R^m$, $t \in I$, $y \in Y$ and $z \in Z$; k is a Lebesgue measurable function on I such that

$$\int_0^1 k(t)dt = A < \infty . \tag{3}$$

Under conditions (2) and (3), equation (1) has a unique solution $x(t)$ corresponding to any given *initial condition* $x(0) = x_o$. We consider, for the time being, that the initial condition $x(0) = 0$; the resulting solution $x(t)$ is called the *trajectory* corresponding to $(y(t), z(t))$. We then compute a *"payoff"* $P(y,z)$ given by

$$P(y,z) = \mu(x(t)) + \int_0^1 h(t, x(t), y(t), z(t))\,dt. \tag{4}$$

Here $h\colon I \times R^m \times Y \times Z \to R$ is continuous and μ is a continuous real-valued function on the Banach space $[C(I)]^m$ of continuous function $x\colon I \to R^m$.

The game is "zero-sum" so that the object of player J_1 is to maximize the payoff P, while J_2 aims to minimize P. At each time t, player J_1 is aware of the complete history of the game from time 0 to time t, while J_2 is in a similar position.

2. THE CONCEPT OF VALUE

We denote by M_1 the set of all measurable functions $y\colon I \to Y$ (in the sense of §1), modulo the identification of any two functions equal almost everywhere; M_2 is similarly defined with Z replacing Y. We need the following lemma for future reference; this explains our restriction to metrizable spaces Y and Z.

LEMMA 2.1 *Two measurable functions* $y\colon I \to Y$ *and* $y'\colon I \to Y$ *are equal almost everywhere if and only if for every continuous function* $\varphi\colon Y \to R$

$$\varphi(y(t)) = \varphi(y'(t)) \qquad \text{a.e.}$$

PROOF

Let d be a metric on Y describing the topology. As Y is compact and metrizable, Y has a countable dense subset $\{\eta_k;\ k = 1,2,\dots\}$; consider the functions

$$\varphi_k(\eta) = d(\eta,\eta_k).$$

By assumption for each k

$$\varphi_k(y(t)) = \varphi_k(y'(t)) \qquad \text{a.e.}$$

and so

$$\varphi_k(y(t)) = \varphi_k(y'(t)),\quad k = 1,2,\dots, \qquad \text{a.e.}$$

It follows easily that, almost everywhere,

$$y(t) = y'(t).$$

We are grateful to P. Stefan for the following example. Suppose Y is the non-metrizable space I^I with the product topology; we define two functions $y: I \to I^I$ and $y': I \to I^I$ by

$$\begin{aligned} (y(t))_s &= 0 && \forall\ t \in I,\quad s \in I \\ (y'(t))_s &= 1 && \text{if } t = s \\ &= 0 && \text{if } t \neq s. \end{aligned}$$

By the Stone-Weierstrass theorem, the algebra generated by the co-ordinate maps and the constant functions is dense in $C(I^I)$, and so by expressing $\varphi \in C(I^I)$ as the limit of a sequence of polynomials in these functions, we may show that there exists a countable subset I_φ of I such that

$$\eta_s = \eta'_s \qquad s \in I_\varphi$$

implies

$$\varphi(\eta) = \varphi(\eta').$$

Hence

$$\varphi(y(t)) = \varphi(y'(t) \qquad t \in I - I_\varphi$$

and so

$$\varphi(y(t)) = \varphi(y'(t)) \qquad \text{a.e.}$$

This counter-example shows that metrizability is necessary in Lemma 2.1, and demonstrates our reason for asking that Y and Z be metrizable. However, if we took the conclusion of Lemma 2.1 as the definition of "equal almost everywhere" then we could extend our approach to non-metrizable Y and Z.

The elements of M_1 are called *control functions* for J_1, and similarly M_2 is the set of control functions for J_2. Any map

$$\alpha: M_2 \to M_1$$

is a *pseudo-strategy* for J_1. It prescribes a rule by which J_1 may determine his own control function given J_2's choice of control function. Each pseudo-strategy has a value

$$u(\alpha) = \inf_{z \in M_2} P(\alpha z, z)$$

giving the worst possible result for J_1 if he adopts the pseudo-strategy α. We may similarly define a pseudo-strategy for J_2 as a map $\beta: M_1 \to M_2$ and its value is given by

$$v(\beta) = \sup_{y \in M_1} P(y, \beta y).$$

In practice, not all pseudo-strategies are "reasonable", for they imply foreknowledge of the other player's choice of control function. Hence we say $\alpha: M_2 \to M_1$ is a *strategy* if whenever $0 < T \leq 1$ and

$$z_1(t) = z_2(t) \quad \text{a.e.} \quad 0 \leq t \leq T$$

then

$$\alpha z_1(t) = \alpha z_2(t) \quad \text{a.e.} \quad 0 \le t \le T. \tag{5}$$

We make a similar definition of strategy for J_2. The set of all strategies for J_1 is denoted by Γ, and the set of strategies for J_2 by Δ.

The *value* of the game to J_1 is the best he can force by adopting a strategy, i.e.

$$U = \sup_{\alpha \in \Gamma} u(\alpha)$$

while the value to J_2 is

$$V = \inf_{\beta \in \Delta} v(\beta).$$

If $U = V$, then the game has a value, in the sense that neither player can force a better result than V, and both players can (almost) force V. In general it is not true that $U = V$, and the following classical example of Berkovitz [2] precisely demonstrates the reasons for this and the defects of this definition.

Example 2.2 Let $Y = Z = [-1,1]$ and suppose the dynamics of G are given by

$$\frac{dx}{dt} = (y-z)^2$$

where $x \in R$. Let the payoff be

$$P = \int_0^1 x(t)\,dt.$$

It is easy to see that a 'best' strategy α for J_1 is given by

$$\alpha z(t) = 1 \quad \text{if} \quad z(t) < 0$$
$$= -1 \quad \text{if} \quad z(t) \geq 0 ,$$

while a best strategy β for J_2 is given by

$$\beta y(t) = y(t) .$$

Then

$$U = u(\alpha)$$
$$= P(\alpha z_o, z_o)$$

where

$$z_o(t) \equiv 0 .$$

Hence

$$U = \int_0^1 t\,dt$$
$$= \tfrac{1}{2} .$$

Similarly

$$V = v(\beta)$$
$$= 0 ,$$

and $U \neq V$.

A further problem is that α and β cannot in any reasonable sense be played against each other. If J_1 elects to adopt strategy α, while J_2 adopts β then we cannot determine any outcome to the game; there are no control functions y and z with

$$\alpha z = y$$
$$\beta y = z .$$

3. THE VALUE IN THE SENSE OF FRIEDMAN

In this section we describe an alternative definition of value used by Friedman ([10]) and relate it to our own definition.

Let N be an integer and let $\delta = 2^{-N}$; we shall define a game E_N^+. Let

$$I_1 = [0,\delta]$$

$$I_j = ((j-1)\delta\,,\, j\delta] \qquad j = 2,3,\dots,2^N.$$

The game E_N^+ has the same dynamics, initial condition and payoff as G (given by equations (1)-(4)) but is played in the following manner: J_2 first selects his control function on I_1 and then J_1 selects his control function on I_1, and the players then play alternately, J_2 selecting his function on I_j before J_1 selects his function on I_j at the jth step.

Let $M_1^{(j)}$ and $M_2^{(j)}$ denote the spaces of measurable functions $I_j \to Y$ and $I_j \to Z$ respectively, in which, as before, two functions equal almost everywhere are identified (Friedman [10] does not make this identification, but this makes no difference to the discussion). A strategy for J_1 in the game E_N^+ is then a collection of maps

$$\Sigma = (\Sigma_1, \dots \Sigma_{2^N})$$

where

$$\Sigma_j\colon M_2^{(1)} \times \dots \times M_2^{(j)} \to M_1^{(j)}$$

and a strategy Π for J_2 in E_N^+ is an element z_1 of $M_2^{(1)}$ together with a collection of maps $(\Pi_2, \dots, \Pi_{2^N})$

$$\Pi_j\colon M_1^{(1)} \times \dots \times M_1^{(j-1)} \to M_2^{(j)}.$$

Σ then determines a rule of procedure for J_1 and Π determines a rule for J_2. Given two such rules (Σ, Π) there is an outcome to the game (i.e. a pair of control functions determined by Σ and Π) and a payoff $P(\Sigma, \Pi)$; it follows from the theory of alternate move games that

$$\inf_{\Pi} \sup_{\Sigma} P(\Sigma,\Pi) = \sup_{\Sigma} \inf_{\Pi} P(\Sigma, \Pi)$$

$$= V_N^+, \text{ say.}$$

Then the game E_N^+ has value V_N^+ (see Friedman [10] for the details of this approach).

It follows quickly that $V_N^+ \geq V_{N+1}^+$ for all N and so we may define

$$V^+ = \lim_{N\to\infty} V_N^+ .$$

V^+ is the *upper value* of G in the sense of Friedman.

If we define E_N^- as the game played as E_N^+ except that at each step J_1 plays first, we may show E_N^- has a value V_N^- and define

$$V^- = \lim_{N\to\infty} V_N^- ,$$

the *lower value* of G.

We have immediately

$$V_N^- \leq V_N^+ \quad \text{for all } N$$

and so

$$V^- \leq V^+ .$$

We say G has value in the sense of Friedman if $V^- = V^+$.

We next present a re-interpretation of the Friedman values V^- and V^+

in terms of pseudo-strategies. For $-1 \leq s \leq 1$, we define $\Gamma(s)$ as the set of pseudo-strategies α for J_1 such that whenever

$$z_1(t) = z_2(t) \quad \text{a.e.} \quad 0 \leq t \leq T, \quad \text{where} \quad T > 0,$$

then

$$\alpha z_1(t) = \alpha z_2(t) \quad \text{a.e.} \quad 0 \leq t \leq \min(T+s, 1). \tag{6}$$

We define $\Delta(s)$ similarly for player J_2. Thus $\Gamma(s)$ is the set of pseudo-strategies available to J_1 if he has a reaction time s (which may be negative, in which case he is 'anticipating' his opponent's moves).

Let

$$U(s) = \sup_{\alpha \in \Gamma(s)} u(\alpha)$$

$$V(s) = \inf_{\beta \in \Delta(s)} v(\beta).$$

Then U and V are monotone functions of s and so we may define

$$U^+(s) = \lim_{t \to s+} U(t)$$

$$U^-(s) = \lim_{t \to s-} U(t)$$

$$V^+(s) = \lim_{t \to s+} V(t)$$

$$V^-(s) = \lim_{t \to s-} V(t).$$

We shall show that $V^+(0) = V^+$ and $V^-(0) = V^-$ later.

THEOREM 3.1 *For all* s, $-1 \leq s \leq 1$,

$$V^+(s) = U^-(-s),$$

$$V^-(s) = U^+(-s).$$

PROOF

Let $s = \frac{p}{2^N}$ be dyadically rational and positive; we consider a game $E_N^+(s)$ played thus: the dynamics, initial condition and payoff are as in E_N^+, but now J_2 selects his control function initially on $I_1 \cup \ldots \cup I_{p+1}$, and then J_1 selects his function on I_1, then J_2 on I_{p+2}, etc. until J_2 finally selects his control function on I_{2^N} and then J_1 on $I_{2^N - p} \cup \ldots \cup I_{2^N}$. As before $E_N^+(s)$ has value which we denote by $V_N^+(s)$.

A strategy for J_1 in $E_N^+(s)$ is given by a collection of maps $\Sigma = (\Sigma_1, \ldots, \Sigma_{2^N - p})$ where

$$\Sigma_j : M_2^{(1)} \times \ldots \times M_2^{(j+p)} \to M_1^{(j)} .$$

Clearly Σ determines a pseudo-strategy α for J_1 and $\alpha \in \Gamma(-s-\delta)$, where $\delta = 2^{-N}$; conversely if $\alpha \in \Gamma(-s)$, we may determine a strategy Σ in $E_N^+(s)$ corresponding to α. By this reasoning we conclude that

$$U(-s-\delta) \geq V_N^+(s) \geq U(-s),$$

and a similar argument yields

$$V(s+\delta) \geq V_N^+(s) \geq V(s).$$

Let us now assume that s is any real number $0 \leq s \leq 1$, and that $s_n \downarrow s$ is a sequence of dyadic rationals decreasing to s; let $s_n = \frac{p_n}{2^n}$.

Then we have

$$U^-(-s) = \lim_{n \to \infty} V_n^+(s_n)$$

and

$$V^+(s) = \lim_{n \to \infty} V_n^+(s_n)$$

so that

$$U^-(-s) = V^+(s).$$

We similarly obtain for $0 < s \leq 1$ that

$$U^+(-s) = V^-(s)$$

and by the same method, we obtain the same relationships for $s \leq 0$ (it is necessary to consider a game like $E_N^+(s)$ in which J_1 plays first).

Before establishing that $V^+(0) = V^+$, we require a technical lemma establishing bounds on the possible trajectories in the game G.

LEMMA 3.2 *There exists a constant* $L < \infty$ *such that for any trajectory* $x(t)$ *in* G

$$\sup_{t \in I} \|x(t)\| \leq L.$$

PROOF

Since f is continuous and $I \times Y \times Z$ is compact we have

$$\sup_{t \in I} \sup_{y \in Y} \sup_{z \in Z} \|f(t,0,y,z)\| = B < \infty.$$

Let $x(t)$ be trajectory corresponding to the control functions $(y(t), z(t))$; then we have

$$\|x(t)\| \leq \int_0^t \|\dot{x}(t)\| ds$$

$$= \int_0^t \|f(s,x(s),y(s),z(s))\| ds$$

$$\leq Bt + \int_0^t k(s)\|x(s)\| ds$$

by equation (2).

Let

$$A(t) = \int_0^t k(s)ds.$$

Then

$$Bte^{A(t)} = \int_0^t (Be^{A(s)} + Bsk(s)e^{A(s)})ds$$

$$\geq Bt + \int_0^t Bsk(s)e^{A(s)}ds$$

and hence

$$Bte^{A(t)} - \|x(t)\| \geq \int_0^t k(s)(Bse^{A(s)} - \|x(s)\|)ds.$$

Now let

$$\varphi(t) = Bte^{A(t)} - \|x(t)\| ;$$

we have that φ is continuous and $\varphi(0) = 0$. Let

$$T = \sup(t: \varphi(s) \geq 0 , \ 0 \leq s \leq t)$$

and suppose $T < 1$. Choose δ_1 such that $T + \delta_1 < 1$ and

$$0 \leq \int_T^{T+\delta_1} k(s)ds < 1 .$$

For $T \leq t \leq T + \delta_1$

$$\varphi(t) \geq \int_0^t k(s)\varphi(s)\,ds$$

$$\geq \int_T^t k(s)\varphi(s)\,ds$$

$$\geq \gamma \int_T^t k(s)\,ds$$

where

$$\gamma = \inf\{\varphi(s); \ T \leq s \leq t\} < 0 .$$

Hence

$$\varphi(t) \geq \gamma \int_T^{T+\delta_1} k(s)\,ds$$

and

$$\gamma \geq \gamma \int_T^{T+\delta_1} k(s)\,ds .$$

This contradicts the fact that $\gamma < 0$, and so we have $T = 1$, i.e.

$$\|x(t)\| \leq Bte^{A(t)}$$

$$\leq Be^{A}$$

for all t.

LEMMA 3.3 *The set of all trajectories in* G *is relatively compact in the Banach space* $[C(I)]^m$ *of continuous functions* $x: I \to R^m$; *the functional* μ *is uniformly continuous on the set of trajectories.*

PROOF

As f is continuous on the compact set $C = \{\|x\| \leq L\} \times I \times Y \times Z$ we have

$$\sup_{(t,x,y,z)\in C} \|f(t,x,y,z)\| = B' < \infty .$$

For any trajectory $x(t)$, by Lemma 3.2,

$$\|x(t) - x(s)\| \leq B'|t-s|$$

where $1 \geq t > s \geq 0$. Thus the trajectories are equicontinuous and bounded, and so relatively compact in $[C(I)]^m$ by the Ascoli-Arzela Theorem. It follows immediately that μ is uniformly continuous on the trajectories.

THEOREM 3.4

$$V^+ = V^+(0) = U^-(0) ,$$

and

$$V^- = V^-(0) = U^+(0) .$$

PROOF

Let $\delta = 2^{-N}$ for $N \geq 1$ and consider the game E_N^+ (with value V_N^+); a pseudo-strategy β in $\Delta(\delta)$ for J_2 induces as in Theorem 3.1 a strategy Π in E_N^+, and we conclude that

$$V_N^+ \leq V(\delta) ,$$

and so letting $\delta \to 0$

$$V^+ \leq V^+(0) . \tag{7}$$

Conversely suppose $\varepsilon > 0$; we shall show that there exists a strategy β^* for J_2 in $\cup(\Delta(s);\ s > 0)$ such that

$$v(\beta^*) \leq V^+ + \varepsilon .$$

Let L, B' and $A(t)$ be defined as above; in addition let

$$h = \sup_{\|x\| \leq L} \sup_{t \in I} \sup_{y \in Y} \sup_{z \in Z} |h(t,x,y,z)| ,$$

and choose $\eta > 0$ such that if $\|x\| \leq L$, $\|x'\| \leq L$ and $\|x - x'\| \leq \eta$ then

$$|h(t,x,y,z) - h(t,x',y,z)| \leq \varepsilon/5 , \tag{8}$$

for $(t,y,z) \in I \times Y \times Z$ (using uniform continuity of h). We also assume η is small enough so that if $x(t)$ and $x'(t)$ are any two trajectories with

$$\sup_{t \in I} \|x(t) - x'(t)\| \leq \eta$$

then

$$|\mu(x(t)) - \mu(x'(t))| \leq \varepsilon/5 , \tag{9}$$

(using Lemma 3.3).

Choose $\delta = 2^{-N}$ such that

$$V_N^+ - V^+ \leq \varepsilon/5$$

and a strategy Π for J_2 in E_N^+ such that

$$\sup_y P(y, \Pi y) \leq V_N^+ + \varepsilon/5 .$$

This induces a strategy β in G for J_2 with

$$v(\beta) \leq V^+ + 2\varepsilon/5 . \tag{10}$$

Let z_0 be a fixed member of Z and suppose $\eta_0 < 1$, and

$$2h\eta_0 \leq \varepsilon/5 , \tag{11}$$

$$\eta_0 \leq \frac{\eta e^{-A}}{2B} . \tag{12}$$

We define β^* by

$$\beta^* y(0) = z_0 ,$$

$$\beta^* y(t) = z_0 , \; j\delta < t \leq (j + \eta_0)\delta ,$$

for $j = 0,1,2,\ldots,2^N - 1$, and

$$\beta^* y(t) = \beta y(t) , \quad (j+\eta_0)\delta < t \leq (j+1)\delta , \tag{13}$$

for $j = 0,1,2,\ldots 2^N - 1$.

Then it is clear that $\beta^* \in \Delta(\eta_0 \delta)$,

and so

$$V^+(0) \leq v(\beta^*) .$$

Let $y(t)$ be a control function for J_1 and let $x(t)$ and $x^*(t)$ be the trajectories corresponding to $(y, \beta y)$ and $(y, \beta^* y)$; then we have

$$\frac{d}{dt}(x^*(t) - x(t)) = f(t,x^*(t),y(t),\beta^* y(t)) - f(t,x(t),y(t),\beta y(t))$$

and so from the definition of β^* (13) and the Lipschitz condition (2) satisfied by f we have

$$\|x^*(t) - x(t)\| \leq \theta(t) ,$$

where $\theta(t)$ satisfies the differential equation

$$\frac{d\theta}{dt} = k(t)\theta + 2B'\chi(t) \tag{14}$$

Here χ is the characteristic function of the set $\bigcup_{j>0} [j\delta, (j+\eta_o)\delta]$,

and $\theta(0) = 0$. Hence we obtain, solving (14),

$$\theta(t) = 2B'e^{A(t)}\int_0^t e^{-A(s)}\chi(s)\,ds ,$$

and so, $|\theta(t)| \leq 2B'e^A(\int_0^1 \chi(s)\,ds)$ for $0 \leq t \leq 1$.

$$= 2B'e^A\eta_o \quad \text{by (12)}.$$

Therefore

$$\|x^*(t) - x(t)\| \leq \eta$$

and so, by (9)

$$|\mu(x^*(t)) - \mu(x(t))| \leq \varepsilon/5 . \tag{15}$$

Furthermore, by (8),

$$\int_0^1 |h(t,x^*,y,\beta^*y) - h(t,x,y,\beta y)|\,dt$$

$$\leq \sum_{j=0}^{2^N-1} \{2h\eta_o\delta + \int_{(j+\eta_o)\delta}^{(j+1)\delta} |h(t,x^*,y,\beta y) - h(t,x,y,\beta y)|\,dt\}$$

$$\leq 2h\eta_o + (1-\eta_o)\varepsilon/5$$

$$\leq 2\varepsilon/5 . \tag{16}$$

Combining (15) and (16)

$$|P(\beta y,y) - P(\beta^* y,y)| \leq 2\varepsilon/5 + \varepsilon/5 = 3\varepsilon/5 ,$$

and therefore

$$|v(\beta) - v(\beta^*)| \leq 3\varepsilon/5 ,$$

and so $$v(\beta^*) \leq V^+ + \varepsilon \qquad \text{by (10)}.$$

i.e. $$V^+(0) \leq V^+ + \varepsilon$$

and so $$V^+(0) = V^+, \qquad \text{by (7)}.$$

COROLLARY. $$V^- \leq U, \quad V \leq V^+.$$

Theorem 3.1 and Theorem 3.4 raise several issues of interest. We see that V^+ is to be interpreted as the value of the game G to the player J_1 if he is, in some sense, almost able to anticipate J_2's play in the future (for $V^+ = U^-(0)$). The value V is the value to J_1 provided that his reactions are instantaneous. The smallest value V^- is the realistic value of the game to J_1; it is obtained by giving J_1 a reaction time and letting this reaction tend to zero. V^- is the only value appropriate to J_1 unless he is assumed to be superhuman.

One might ask whether it is reasonable to hope for a value if we assume in G that each player has a certain reaction time; suppose J_1 has reaction time $s_1 > 0$ and J_2 has reaction time $s_2 \geq 0$. The existence of value requires that

$$U(s_1) = V(s_2)$$

(both in the sense described in §2 or in the sense of Friedman) so that for $0 < s < s_1$ we have

$$\begin{aligned} U(s_1) &\geq U^+(s) \\ &= V^-(-s) \\ &\geq V(s_2) \end{aligned}$$

and hence

$$U(s_1) = U^+(s)$$

for $0 < s \leq s_1$. In particular

$$U(s_1) = V^-$$

so that, if this game, with reaction time, has value, then player J_1 does not lose ground by having a reaction time. This can clearly only happen in pathological situations. In particular, if we take $s_1 > 0$ and $s_2 = 0$ we may interpret this as follows: while J_1 obviously loses ground by having a positive reaction time (i.e. $U(s_1) < U(0)$), J_2 cannot exploit his advantage, for he can never vary from his best strategy in G in case J_1, by accident, hits upon the correct line of defence. For this reason, we are unable to understand precisely what is intended in §9 of [10].

Let us now consider a variation of G in which J_1 selects his control function on the interval $-s \leq t \leq 1-s$ (where $s > 0$) and the dynamics are given by

$$\frac{dx}{dt} = f(t, x(t), y(t-s), z(t)), \tag{17}$$

subject to the initial condition $x(0) = 0$ and the payoff is given by (4). This game is effectively G subject to the restriction that J_1 has a reaction time s, *and* J_2 has a negative reaction time $-s$. For at time t_o, J_2 is not only aware of the value of $y(t_o - s)$ but also of $y(t-s)$ for $t_o \leq t \leq t_o + s$. This game has value if

$$U(s) = V(-s).$$

As U and V are monotone functions their discontinuities are countable, and so we conclude that except for a countable number of values of s

$$\begin{aligned} U^+(s) &= U^-(s) \\ &= V^+(-s) \\ &= V^-(-s) \end{aligned}$$

and so

$$U(s) = V(-s) .$$

Thus this variation of G has a value for almost every choice of s ; we conjecture that U and V are continuous for $s \neq 0$ and so $U(s) = V(-s)$ for $s \neq 0$.

4. ANOTHER FORMULA FOR THE FRIEDMAN UPPER AND LOWER VALUES

In this section we show the Friedman values V^+ and V^- of the game G may be obtained by considering discrete versions H_N^+ and H_N^- of the games E_N^+ and E_N^-. We shall assume throughout this section that the payoff function (equation (4)) takes the special form

$$P(y,z) = g(x(1)) + \int_0^1 h(t,x(t),y(t),z(t))\,dt \tag{18}$$

where $g: R^m \to R$ is a continuous function; that is we assume

$$\mu(x(t)) = g(x(1)) .$$

We now define the game H_N^+. Let N be a positive integer and let $\delta = 2^{-N}$; the players J_1 and J_2 select their control functions alternately on the intervals I_j , $j = 1,2,\dots,2^N$ as in E_N^+ (see description at the beginning of §3). As in E_N^+, J_2 plays first at each step. The dynamics of H_N^+ are given by

$$x(0) = 0$$

$$x(t_j) = x(t_{j-1}) + \int_{t_{j-1}}^{t_j} f(t_{j-1},x(t_{j-1}),y(t),z(t))\,dt \tag{19}$$

where

$$t_j = j\delta \qquad 0 \leq j \leq 2^N .$$

The payoff is given by

$$P(y(t),z(t)) = g(x(1)) + \sum_{j=1}^{2^N} \int_{t_{j-1}}^{t_j} h(t_{j-1}, x(t_{j-1}), y(t), z(t))\,dt \qquad (20)$$

(We recall that G is assumed to have a payoff given by (18).)

Once again we use the general theory of alternate move games to deduce that the game H_N^+ has a value which we denote by S_N^+.

THEOREM 4.1 $$\lim_{N \to \infty} S_N^+ = V^+ .$$

PROOF

We shall in fact show that

$$\lim_{n \to \infty} (V_N^+ - S_N^+) = 0 .$$

Suppose $\varepsilon > 0$; then in order to show that

$$|V_N^+ - S_N^+| \leq \varepsilon$$

it is only necessary to show that for any pair of control functions $(y(t), z(t))$

$$|P_E(y,z) - P_H(y,z)| \leq \varepsilon$$

where $P_E(y,z)$ denotes the payoff in E_H^+ corresponding to $(y(t),z(t))$ and $P_H(y,z)$ is the payoff in H_N^+. This is readily seen since the method of selection of control functions is identical in the two games, and so for any pair of strategies (Σ, Π) for E_N^+ or H_N^+ we should then have

$$|P_E(\Sigma, \Pi) - P_H(\Sigma, \Pi)| \leq \varepsilon .$$

From this it follows easily that $|V_N^+ - S_N^+| \leq \varepsilon$. We shall apply this principle again to relate values of different games later on.

Let B' be as in Lemma 3.3 and suppose that $D > 0$ is any constant; as f is continuous, it is uniformly continuous on the set $I \times (\|x\| \leq D + L) \times Y \times Z$. Therefore let

$$\rho(\delta) = \sup_{|t' - t| \leq \delta} \sup_{\|x\| \leq D+L} \sup_{y} \sup_{z} \|f(t',x,y,z) - f(t,x,y,z)\|$$

so we have

$$\lim_{\delta \to 0} \rho(\delta) = 0.$$

Now fix $\delta = 2^{-N}$ such that

$$\delta \leq \tfrac{1}{2} e^{-A} \left(\frac{D}{AB'}\right) \tag{21}$$

and

$$\rho(\delta) \leq \tfrac{1}{2} e^{-A} D \tag{22}$$

Suppose $(y(t), z(t))$ is any pair of control functions for J_1 and J_2 and let $x(t)$ be the trajectory corresponding to them in E_N^+ (or G, as G has the same dynamics) and let $\tilde{x}(t)$ be the trajectory in H_N^+. Let

$$\xi_j = \|x(t_j) - \tilde{x}(t_j)\|, \qquad j = 0,1,2,\ldots 2^N$$

so

$$\xi_0 = 0.$$

Provided $\xi_{j-1} \leq D$, we have

$$\|\tilde{x}(t_{j-1})\| \leq \|x(t_{j-1})\| + D$$
$$\leq L + D$$

and so

$$\|f(t,\tilde{x}(t_{j-1}),y(t),z(t)) - f(t_{j-1}\tilde{x}(t_{j-1}),y(t),z(t)\|$$
$$\leq \rho(\delta)$$
$$\|f(t,x(t),y(t),z(t)) - f(t,\tilde{x}(t_{j-1}),y(t),z(t))\|$$
$$\leq k(t)\|x(t) - \tilde{x}(t_{j-1})\|$$
$$\leq k(t)(\xi_{j-1} + B'\delta).$$

Hence

$$|\xi_j - \xi_{j-1}| \leq \int_{t_{j-1}}^{t_j} \|f(t,x(t),y(t),z(t)) - f(t_{j-1},x(t_{j-1}),y(t),z(t))\|dt$$

$$\leq \delta\rho(\delta) + (\xi_{j-1} + B'\delta)\int_{t_{j-1}}^{t_j} k(t)dt$$

$$= \delta\rho(\delta) + (\xi_{j-1} + B'\delta)a_j$$

where

$$a_j = \int_{t_{j-1}}^{t_j} k(t)dt = A(t_j) - A(t_{j-1}).$$

Provided $\xi_1,\xi_2,\ldots,\xi_{j-1} \leq D$ we have

$$\xi_j \leq (1+a_j)\xi_{j-1} + a_jB'\delta + \delta\rho(\delta)$$

$$\leq (1+a_j)(1+a_{j-1})\xi_{j-2} + (1+a_j)(a_{j-1}B'\delta + \delta\rho(\delta)) + a_jB'\delta + \delta\rho(\delta)$$

$$\leq \sum_{i=1}^{j-1} (1+a_j) \ldots (1+a_{i+1})(a_iB'\delta + \delta\rho(\delta)) + a_jB'\delta + \delta\rho(\delta).$$

But

$$(1+a_j) \ldots (1+a_{i+1}) \leq \prod_{i=1}^{2^N} (1+a_i)$$

$$\leq \exp\left(\sum_{i=1}^{2^N} a_i\right)$$

$$= e^A$$

and so

$$\xi_j \leq e^A \sum_{i=1}^{j} (a_i B'\delta + \delta\rho(\delta))$$

$$\leq e^A(AB'\delta + \rho(\delta)) .$$

In view of (21) and (22) we see that if $\xi_1, \ldots, \xi_{j-1} \leq D$ then $\xi_j \leq D$, and so by induction we have

$$\max_{1 \leq j \leq 2^N} \xi_j \leq e^A(AB'\delta + \rho(\delta)).$$

As g and h are continuous functions they are uniformly continuous on the sets $\{\|x\| \leq L + D\}$ and $I \times \{\|x\| \leq L + D\} \times Y \times Z$ respectively. Then, given $\varepsilon > 0$ we may choose $\delta(\varepsilon) = 2^{-N}$ to satisfy (21) and (22) and also

$$|g(x) - g(x')| \leq \varepsilon/2 \tag{23}$$

whenever

$$\|x - x'\| \leq e^A(AB'\delta + \rho(\delta)) + B'\delta$$

and

$$|h(t,x,y,z) - h(t',x',y,z)| \leq \varepsilon/2 , \tag{24}$$

whenever

$$\|x - x'\| \leq e^A(AB'\delta + \rho(\delta)) + B'\delta$$

and

$$|t - t'| \leq \delta .$$

Then for any pair of controls $(y(t),z(t))$ describing trajectories $x(t)$ and $\tilde{x}(t)$ in E_N^+ and H_N^+ we have

$$\|x(t) - \tilde{x}(t_{j-1})\| \leq e^A(AB'\delta + \rho(\delta)) + B'\delta$$

whenever

$$t_{j-1} \leq t \leq t_j ,$$

and so we have by (24)

$$\left| \sum_{j=1}^{2^N} \int_{t_{j-1}}^{t_j} h(t_{j-1},x(t_{j-1}),y(t),z(t))dt - \int_0^1 h(t,x(t),y(t),z(t))dt \right| \leq \varepsilon/2 .$$

Similarly by (11)

$$|g(\tilde{x}(1)) - g(x(1))| \leq \varepsilon/2$$

so that

$$|P_E(y,z) - P_H(y,z)| \leq \varepsilon$$

for N chosen to satisfy (21), (22), (23) and (24).

A similar analysis gives us that

$$\lim_{N\to\infty} S_N^- = V^-$$

where S_N^- is the value of the game H_N^- in which J_1 moves first at each step.

5. THE VALUE IN THE SENSE OF FLEMING

We suppose G has a payoff of the form (18); we now consider the game H_N^+ but impose the further restriction that both players must choose constant functions at each step; this game will be denoted by K_N^+. Thus the players move alternately with J_2 playing first, and at the completion of the game J_1 will have selected a sequence $(y_1,\ldots,y_{2^N})$ of elements of Y and J_2 will have selected a sequence $(z_1,\ldots,z_{2^N})$ of elements of Z. The trajectory will then be determined by $x(0) = 0$ and

$$x(t_j) = x(t_{j-1}) + \delta f(t_{j-1},x(t_{j-1}),y_j,z_j) \qquad (25)$$

and the payoff by

$$P = g(x(1)) + \delta \sum_{j=1}^{2^N} h(t_{j-1}, x(t_{j-1}), y_j, z_j). \tag{26}$$

As usual we can see that K_N^+ will have a value which we denote by W_N^+. Similarly the game K_N^- (in which J_1 plays first at each step) has value W_N^-.

We may also consider K_N^+ starting at time $t_k = k\delta$ and with initial condition $x(t_k) = x$; we thus obtain $K_N^+(t_k,x)$ in which the players select sequences $(y_{k+1},\dots,y_{2^N})$ and $(z_{k+1},\dots,z_{2^N})$ and the trajectory is given by (23) subject to the initial condition $x(t_k) = x$. The payoff in $K_N^+(t_k,x)$ is given by

$$P = g(x(1)) + \delta \sum_{j=k+1}^{2^N} h(t_{j-1}, x(t_{j-1}), y_j, z_j). \tag{27}$$

This game has a value which we denote by $W_N^+(t_k,x)$.

An easy argument from the theory of alternate move games gives us that

$$W_N^+(t_k,x) = \min_{z\in Z} \max_{y\in Y} (W_N^+(t_{k+1},x') + \delta h(t_k,x,y,z)) \tag{28}$$

where

$$x' = x + \delta f(t_k,x,y,z)$$

and

$$W_N^+(1,x) = g(x). \tag{29}$$

Considerations of this kind led Isaacs [13] to derive heuristically the so called Isaacs-Bellman differential equation for the 'upper value' $R(t,x)$ of the game G, subject to the initial conditions

$$x(t) = x$$

and with payoff given by

$$P = g(x(1)) + \int_t^1 h(s,x(s),y(s),z(s))ds \; . \tag{30}$$

This partial differential equation is given by

$$\frac{\partial R}{\partial t} + F^+(t,x,\nabla R) = 0 \tag{31}$$

where

$$F^+(t,x,p) = \min_z \max_y \left(\sum_{i=1}^m p_i f_i(t,x,y,z) + h(t,x,y,z) \right)$$

$$= \min_z \max_y \, (p.f + h) \tag{32}$$

for $p = (p_i) \in R^m$, $x \in R^m$ and $t \in I$. R must also satisfy the boundary condition

$$R(1,x) = g(x) \; . \tag{33}$$

Unfortunately the Isaacs-Bellman equation (31) is highly degenerate and we have no theorems guaranteeing either the existence or uniqueness of solutions of (31). However, Fleming ([5], [6], [7]) has developed an approach circumventing this difficulty to produce a 'reasonable' solution of (31). For Fleming's approach it is necessary to assume conditions (F1) - (F3) where:

(F1) f is uniformly Lipschitz in x, i.e.

$$\sup |k(t)| = k < \infty$$

(where $k(t)$ is the function defined in (2)).

(F2) h is uniformly Lipschitz in x, i.e.

$$|h(t,x_1,y,z) - h(t,x_2,y,z)| \leq M\|x_1 - x_2\|$$

for $t \in I$, $x_1, x_2 \in R^m$, $y \in Y$ and $z \in Z$.

(F3) g is Lipschitz in x, i.e.

$$|g(x_1) - g(x_2)| \le M^* \|x_1 - x_2\| .$$

However, we do not require the full force of Fleming's method and so we impose two further restrictions:

(F4) f and h satisfy uniform Lipschitz condition in t, i.e. there exists $Q > 0$ with

$$\|f(t_1,x,y,z) - f(t_2,x,y,z)\| \le Q|t_1 - t_2|$$

$$|h(t_1,x,y,z) - h(t_2,x,y,z)| \le Q|t_1 - t_2| .$$

(F5) g is twice continuously differentiable and its derivatives $\frac{\partial g}{\partial x_i}, \frac{\partial^2 g}{\partial x_i \partial x_j}$ each satisfy Lipschitz conditions in x.

If G satisfies (F1) - (F5) we shall say that G is of type (F). For games of type (F), Fleming considers the parabolic equation.

$$\frac{\lambda^2}{2} \nabla^2 R + \frac{\partial R}{\partial t} + F^+(t,x,\nabla R) = 0 , \tag{34}$$

subject to $R(1,x) = g(x)$. Quoting results of Friedman ([8] or [9]) or Oleinik and Kruzhkov [15], he observes that (34) has a unique solution $W_\lambda^+(t,x)$ for $\lambda > 0$, and W_λ^+ is continuously differentiable in t and twice continuously differentiable in the space variable x. Furthermore W_λ^+ and its derivatives $\frac{\partial W_\lambda^+}{\partial t}$, $\frac{\partial W_\lambda^+}{\partial x_i}$ and $\frac{\partial^2 W_\lambda^+}{\partial x_i \partial x_j}$ each satisfy Hölder conditions of the form

$$|\psi(t,x) - \psi(t',x')| \le Q'[|t-t'|^{\frac{1}{2}} + |x-x'|] .$$

For $\lambda > 0$ and $\delta = 2^{-N}$ with N an integer, Fleming considers a stochastic difference equation related to (28)

$$W_{N,\lambda}^+(t_j,x) = \min_{z\in Z} \max_{y\in Y} \{\mathscr{E}(W_{N,\lambda}^+(t_{j+1},x') + \delta h(t_j,x,y,z))\} \tag{35}$$

where

$$x' = x + \delta f(t_j, x, y, z) + \delta^{\frac{1}{2}} \lambda \eta_j .$$

Here $(\eta_0 \dots \eta_{2^N - 1})$ is a sequence of normalized mutually independent Gaussian random variables (and $\mathcal{E}$ denotes the *expectation*). $W^+_{N,\lambda}$ is determined for $t_j = j\delta \quad j = 0,1,2,\dots,2^N$ by the boundary condition

$$W^+_{N,\lambda}(1,x) = g(x) .$$

Fleming obtains the following theorems (see [7]).

THEOREM 5.1 $\lim_{N\to\infty} W^+_{N,\lambda}(t,x) = W^+_\lambda(t,x)$ *for* $\lambda > 0$ *and dyadically rational* t, *uniformly on compacta.*

THEOREM 5.2 $\lim_{\lambda\to 0} W^+_{N,\lambda}(t,x) = W^+_N(t,x)$ *uniformly in* N *for each dyadically rational* t, *and* N *such that* $t = p.2^{-N}$ *with* p *an integer.*

From these he deduces:

THEOREM 5.3 $\lim_{\lambda\to 0} W^+_\lambda(t,x) = \lim_{N\to\infty} W^+_N(t,x)$ *for all dyadically rational* t.

In particular $W^+ = \lim_{N\to\infty} W^+_N$ exists. We may also deduce that

$$\lim_{\lambda\to 0} W^+_\lambda(t,x) = W^+(t,x)$$

exists for all $t \in I$ and $x \in R^m$. Fleming shows that the function W^+ is a *generalized* solution of the Isaacs-Bellman equation (31), in the sense that it is a Lipschitz function satisfying (31) almost everywhere.

DEFINITION 5.4 The generalized solution of (31) obtained in this way is called the *Fleming solution* of the Isaacs-Bellman equation (31).

We observe that the Fleming solution of (34) depends only on the function $F^+(t,x,p)$ (and the boundary condition $g(x)$), for it is the limit of the unique solutions of the equations (34) for $\lambda \to 0$.

We may apply the same analysis to the values W_N^- of the games K_N^- and deduce that

$$W^- = \lim_{N\to\infty} W_N^-$$

where $W^- = W^-(0,0)$ and the function $W^-(t,x)$ is the Fleming solution of the equation

$$\frac{\partial R}{\partial t} + F^-(t,x,\nabla R) = 0\,, \tag{36}$$

where

$$F^-(t,x,p) = \max_{y\in Y} \min_{z\in Z} (p.f + h)\,, \tag{37}$$

subject to the boundary condition

$$W^-(1,x) = g(x)\,.$$

If G satisfies (F) and also the *Isaacs condition*

$$F^+(t,x,p) = F^-(t,x,p) \tag{38}$$

for $0 \le t \le 1$, $x \in R^m$ and $p \in R^m$, then as the Fleming solution of (31) or (36) is uniquely determined by the equation, we may deduce

THEOREM 5.5 *If* G *satisfies* (F) *and the Isaacs condition (38) then* $W^+ = W^-$.

6. AN ESTIMATE FOR $W^+ - W^-$

A slightly stronger form of Theorem 5.5 will be required later on, and in this section we develop this result. We wish to show that if $F^+(t,x,p)$ and $F^-(t,x,p)$ are close then $W^+ - W^-$ is small. Once again we assume G satisfies (F).

LEMMA 6.1 *For* $\lambda > 0$ *and* $(t,x) \in I \times R^m$

$$\|\nabla W_\lambda^+(t,x)\| \leq e^{2A}(M + M^*) ,$$

(see (3), (F2) and (F3) for the definitions of A, M, M^*).

PROOF

Let $\delta = \frac{1}{2^N}$ and $t_j = j\delta$ as before; for $\varepsilon < 0$ we choose

$$s_j \in (t_{j-1}, t_j) \quad (j = 1,2,\ldots,2^N)$$

such that

$$k(s_j) \leq k(t) + \varepsilon$$

for $t_{j-1} < t \leq t_j$. Let $s_0 = 0$ and $s_{2^N+1} = 1$, and then define

$$\Omega_N(s_{2^N+1}, x) = g(x)$$

$$\Omega_N(s_j, x) = \min_z \max_y \mathscr{E}(\Omega(s_{j+1}, x') + \delta_j h(s_j, x, y, z))$$

where

$$\delta_j = s_{j+1} - s_j$$

$$x' = x + \delta_j f(s_j, x, y, z) + \lambda \delta_j^{\frac{1}{2}} \eta_j$$

and $\eta_0 \ldots \eta_{2^N}$ is a sequence of mutually independent normalized Gaussian random variables.

Using the method of Theorem 1 of [6] we may show

$$|\Omega_N(s_j, x) - W_\lambda(s_j, x)| = O(\delta)$$

uniformly in x and j, as $\delta \to 0$.

We now show that Ω_N satisfies a Lipschitz condition in x, i.e., for $j \geq 1$

$$|\Omega_N(s_j, x_1) - \Omega_N(s_j, x_2)| \leq M_j \|x_1 - x_2\| \tag{39}$$

where

$$M_j = (M^* + M \sum_{i=j}^{2^N} \delta_i) \exp(\sum_{i=j}^{2^N} \delta_i k(s_i)) .$$

Clearly (39) is valid for $j = 2^N + 1$; we prove the result by induction. Suppose it is true for $j + 1$ and fix $y \in Y$, $z \in Z$, x_1, $x_2 \in R^m$. Let

$$x_1' = x_1 + \delta_j f(s_j, x_1, y, z) + \lambda \delta_j^{\frac{1}{2}} \eta_j$$

$$x_2' = x_2 + \delta_j f(s_j, x_2, y, z) + \lambda \delta_j^{\frac{1}{2}} \eta_j$$

so that

$$\mathscr{E}(\|x_1' - x_2'\|) \leq \|x_1 - x_2\|(1 + \delta_j k(s_j)) .$$

Hence

$$|\mathscr{E}(\Omega_N(s_{j+1}, x_1') + \delta_j h(s_j, x_1, y, z)) - \mathscr{E}(\Omega_N(s_{j+1}, x_2') + \delta_j h(s_j, x_2, y, z)|$$

$$\leq \{(1 + \delta_j k(s_j))M_{j+1} + M\delta_j\}\|x_1 - x_2\|$$

and so we deduce

$$|\Omega_N(s_j, x_1) - \Omega_N(s_j, x_2)| \leq ((1 + \delta_j k(s_j))M_{j+1} + M\delta_j)\|x_1 - x_2\| .$$

However

$$(1 + \delta_j k(s_j))M_{j+1} + M\delta_j \leq \exp(\delta_j k(s_j))M_{j+1} + M\delta_j$$

$$\leq M_j$$

and so the inductive hypothesis (39) is proved.

Next we observe that $\delta_j \leq 2\delta$ and so

$$\sum_{j=1}^{2^N} \delta_j k(s_j) \leq 2 \sum_{j=1}^{2^N} \delta k(s_j)$$

$$\leq 2\int_0^1 (k(t) + \varepsilon)dt$$

$$= 2A + \varepsilon .$$

Now W_λ^+ is a continuous solution of (34). For fixed x_1, x_2 and t we may choose s_j such that $|t-s_j| < \delta$ and then, by uniform continuity in t

$$|W_\lambda^+(t,x_1) - W_\lambda^+(t,x_2)| \le o(\delta) + |W_\lambda^+(s_j,x_1) - W_\lambda^+(s_j,x_2)|$$

$$\le o(\delta) + |\Omega_N(s_j,x_1) - \Omega_N(s_j,x_2)|$$

$$\le (o(\delta) + (M + M^*)e^{2A+\varepsilon})\|x_1 - x_2\| .$$

Letting $\delta \to 0$, as ε is arbitrary we obtain

$$|W_\lambda^+(t,x_1) - W_\lambda^+(t,x_2)| \le (M+M^*)e^{2A}\|x_1 - x_2\|$$

and so we obtain, for all $(t,x) \in I \times R^m$

$$\|\nabla W_\lambda^+(t,x)\| \le (M+M^*)e^{2A}.$$

The following estimate will be used in §9.

LEMMA 6.2 *Suppose* G *satisfies (F1) - (F5) and* f,g,h *each vanish outside some compact set. If, whenever*

$$\|p\| \le e^{2A}(M+M^*)$$

we have

$$|F^+(t,x,p) - F^-(t,x,p)| \le \varepsilon$$

then

$$W_\lambda^+(t,x) - W_\lambda^-(t,x) \le e\varepsilon$$

for $\lambda > 0$ *and* $(t,x) \in I \times R^m$. *Consequently*

$$W^+ - W^- \le e\varepsilon .$$

PROOF

It is easy to show (e.g. by using the difference equation for $W_{N,\lambda}^+$) that as $\|x\| \to \infty$ $W_\lambda^+(t,x) \to 0$, and $W_\lambda^-(t,x) \to 0$, since f, g and h vanish outside a compact set.

Consider

$$\theta(t,x) = e^{t}(W_{\lambda}^{+}(t,x) - W_{\lambda}^{-}(t,x)) .$$

Then

$$\theta(1,x) = 0$$

for $x \in R^m$, and $\theta(t,x) \to 0$ uniformly in t as $\|x\| \to \infty$. Hence if $\theta(t,x) > 0$ for some t, x then θ must take its maximum at some (t_o,x_o) where $0 \le t_o < 1$. Clearly

$$\frac{\lambda^2}{2} \nabla^2 \theta(t_o,x_o) + \frac{\partial\theta}{\partial t}(t_o,x_o) \le 0$$

and

$$\nabla\theta(t_o,x_o) = 0.$$

However

$$\nabla\theta = e^{t}(\nabla W_{\lambda}^{+} - \nabla W_{\lambda}^{-})$$

so that

$$\nabla W_{\lambda}^{+}(t_o,x_o) = \nabla W_{\lambda}^{-}(t_o,x_o)$$

$$= p_o, \text{ say},$$

where

$$\|p_o\| \le e^{2A}(M + M^*)$$

by Lemma 6.1.

Hence

$$\frac{\lambda^2}{2} \nabla^2\theta + \frac{\partial\theta}{\partial t} = e^{+t}\{F^{-}(t,x,\nabla W^{-}) - F^{+}(t,x,\nabla W^{+})\} + \theta$$

so that at (t_o,x_o) we have

$$\theta(t_o,x_o) \le e^{t_o}(F^{+}(t_o,x_o,p_o) - F^{-}(t_o,x_o,p_o))$$

$$\le \varepsilon e^{t_o} .$$

Hence for all $(t,x) \in I \times R^m$

$$\theta(t,x) \le e\varepsilon$$

$$W_{\lambda}^{+}(t,x) - W_{\lambda}^{-}(t,x) \le e\varepsilon .$$

The last part follows, letting $\lambda \to 0$.

7. RELAXED CONTROLS

Relaxed controls were introduced into control theory in [20] and into differential game theory in [3]; we refer the reader to [3] for a more detailed discussion.

We denote by $\Lambda(Y)$ and $\Lambda(Z)$ the sets of regular probability measures on Y and Z. By the Riesz representation theorem we may consider $\Lambda(Y)$ as a subset of $C(Y)^*$ and thus give it the weak*-topology $\sigma(C(Y)^*, C(Y))$. If Y is metrizable, then $C(Y)$ is separable and $\Lambda(Y)$ is a compact metrizable space. We may further identify Y as a closed subset of $\Lambda(Y)$ by identifying $y \in Y$ with the probability measure δ_y concentrated at y. We treat $\Lambda(Z)$ in a similar fashion and then extend the definitions of f and h thus:

$$f\colon I \times R^m \times \Lambda(Y) \times \Lambda(Z) \to R^m$$

$$f_i(t,x,\sigma,\tau) = \int_Z \int_Y f_i(t,x,y,z)\, d\sigma(y)\, d\tau(z) \qquad (40)$$

$$i = 1,2,\ldots,m$$

$$h\colon I \times R^m \times \Lambda(Y) \times \Lambda(Z) \to R$$

$$h(t,x,\sigma,\tau) = \int_Z \int_Y h(t,x,y,z)\, d\sigma(y)\, d\tau(z) \qquad (41)$$

(The order of integration is immaterial by Fubini's Theorem.)

It is easy to verify that the extended f and h will satisfy Lipschitz and continuity conditions of the same type as satisfied by the original f and h. We may therefore consider four versions of G, which we denote by G, G_1, G_2, G_{12}. G is the original game described in §1, while in G_1 J_1 is allowed the use of relaxed controls, i.e. he may choose $\sigma \in \Lambda(Y)$ at each instant of time; in G_2, J_2 may use relaxed controls, and in G_{12}, both players may use relaxed controls. We may treat all four games as in the preceeding discussion (§1 - 6); henceforward we denote by a subscript 1, 2 or 12, that a particular quantity or object refers to the game G_1, G_2 or G_{12}.

One great advantage of relaxed controls is demonstrated by

THEOREM 7.1 *If* G *satisfies (F1)-(F5) then*

(i) $W_1^+ = W^+$

(ii) $W_2^- = W^-$

(iii) $W_{12}^+ = W_{12}^- = W_2^+ = W_1^-$.

PROOF

(i) $W^+ = W^+(0,0)$ where $W^+(t,x)$ is the Fleming solution of the Isaacs-Bellman equation (31), subject to $W^+(1,x) = g(x)$. Similarly $W_1^+ = W_1^+(0,0)$ where $W_1^+(t,x)$ is the Fleming solution of

$$\frac{\partial R}{\partial t} + F_1^+(t,x,\nabla R) = 0 \tag{42}$$

subject to $W_1^+(1,x) = g(x)$,

where

$$F_1^+(t,x,p) = \min_{z\in Z} \max_{\sigma\in\Lambda(Y)} (p.f + h).$$

Thus as the Fleming solution of the equation (31) or (42) is uniquely determined by the equation (see Definition 5.4 and the remarks preceeding Theorem 5.5) it is enough to show that

$$F_1^+(t,x,p) = F^+(t,x,p)$$

for $t \in I$, $x \in R^m$ and $p \in R^m$.

Now clearly

$$F_1^+(t,x,p) \geq \min_{z \in Z} \max_{y \in Y} (p.f + h)$$

$$= F^+(t,x,p).$$

However for fixed $z \in Z$

$$\sum_{i=1}^m p_i f_i(t,x,\sigma,z) + h(t,x,\sigma,z) = \int_Y \sum_{i=1}^m p_i f_i(t,x,y,z) + h(t,x,y,z)\, d\sigma(y)$$

by (40) - (41)

$$\leq \max_{y \in Y} \left(\sum_{i=1}^m p_i f_i(t,x,y,z) + h(t,x,y,z) \right)$$

so that

$$F_1^+(t,x,p) \leq F^+(t,x,p),$$

and so

$$F_1^+(t,x,p) \equiv F^+(t,x,p).$$

(ii) is proved similarly.

(iii): We have

$$F_{12}^+(t,x,p) = \min_{\tau \in \Lambda(Z)} \max_{\sigma \in \Lambda(Y)} (p.f + h)$$

$$F_{12}^-(t,x,p) = \max_{\sigma \in \Lambda(Y)} \min_{\tau \in \Lambda(Z)} (p.f + h).$$

For fixed $(t,x,p) \in I \times R^m \times R^m$, $p.f + h$ is a continuous function on $Y \times Z$ and so we may use a well-known result from game theory, due to Wald [19], that

$$F_{12}^+(t,x,p) = F_{12}^-(t,x,p)$$

and hence deduce as in (i)

$$W_{12}^+ = W_{12}^-.$$

If we apply (i) to G_2 we obtain

$$W^+_{12} = W^+_2,$$

while by applying (ii) to G_2, we obtain

$$W^-_{12} = W^-_2.$$

8. THE EXISTENCE OF VALUE UNDER CONDITIONS (F1) - (F5)

In §2 - 5, we introduced six different value concepts for G, i.e. U, V (§2), V^+, V^- (§3) and W^+, W^- (§5); the last two refer only to the case when G satisfies (F1) - (F5) (although they may be defined provided only that G satisfies (F1) - (F3)). Let us note first that as $V^- \leq U$, $V \leq V^+$, we need only consider V^-, V^+, W^- and W^+.

THEOREM 8.1 *If* G *satisfies* (F1) - (F5), *then* $V^+ \leq W^+$ *and* $V^- \geq W^-$.

PROOF

We recall that (Theorem 4.1)

$$V^+ = \lim_{N\to\infty} S^+_N$$

where S^+_N is the value of the game H^+_N. Let us now introduce a variant of H^+_N, called $\overline{H}^+_N$, with the same dynamics, initial condition and pay-off, but with the restriction that at each step J_2 must choose a constant function, while J_1 is still free to choose any function. Clearly $\overline{H}^+_N$ has a value $\overline{S}^+_N$, and is more favourable to J_1, i.e.

$$\overline{S}^+_N \geq S^+_N.$$

We compare $\overline{H}^+_N$ with $K^+_{N,1}$ (see §5; the subscript 1 denotes, as usual, that J_1 may use relaxed controls). For any control function $y(t)$ on I_j, (see

§3), we can define a probability measure σ on Y by

$$\int_Y \varphi \, d\sigma = \delta^{-1} \int_{t_{j-1}}^{t_j} \varphi(y(t))\, dt \tag{43}$$

for $\varphi \in C(Y)$; clearly $\sigma \in \Lambda(Y)$. We now show that if J_2 chooses at the jth step the constant control z, and J_1 replies, in $\overline{H}_N^+$, with the function $y(t)$, then J_1 could achieve the same result by choosing the constant relaxed control σ in $K_{N,1}^+$, where σ is given by (33). For

$$x(t_j) = x(t_{j-1}) + \int_{t_{j-1}}^{t_j} f(t_{j-1}, x(t_{j-1}), y(t), z)\, dt ,$$

(the dynamics of $\overline{H}_N^+$ are given by (19)),

$$= x(t_{j-1}) + \delta \int_Y f(t_{j-1}, x(t_{j-1}), y, z)\, d\sigma(y) \quad \text{(by (43))},$$

$$= x(t_{j-1}) + \delta f(t_{j-1}, x(t_{j-1}), \sigma, z) \quad \text{(by (40))}.$$

Similarly

$$\int_{t_{j-1}}^{t_j} h(t_{j-1}, x(t_{j-1}), y(t), z)\, dt = \delta h(t_{j-1}, x(t_{j-1}), \sigma, z).$$

Thus J_1 can exactly duplicate the effect of any control function in $\overline{H}_N^+$ by a control function in $K_{N,1}^+$ (which is, of course a sequence $\{\sigma_1, \ldots, \sigma_{2N}\}$ of elements of $\Lambda(y)$). In particular, $K_{N,1}^+$ is at least as favourable to J_1 as $\overline{H}_N^+$, i.e.

$$\overline{S}_N^+ \le W_{N,1}^+ .$$

(In fact $\overline{S}_N^+ = W_{N,1}^+$, but this we do not require.)

Thus we have

$$S_N^+ \leq W_{N,1}^+$$

and taking limits

$$V^+ \leq W_1^+ .$$

By Theorem 7.1,

$$V^+ \leq W^+ .$$

The other inequality is a dual result.

THEOREM 8.2 *If* G *satisfies* (F1) - (F5) *and the Isaacs condition (28) i.e.*

$$F^+(t,x,p) = F^-(t,x,p)$$

for all $0 \leq t \leq 1$, $x \in R^m$, $p \in R^m$, *then* $V^+ = V^-$.

PROOF

This follows from Theorems 5.5 and 8.1.

THEOREM 8.3 *If* G *satisfies* (F1) - (F5) *then* $V_{12}^+ = V_{12}^-$.

PROOF

As in Theorem 7.1, we observe that G_{12} satisfies the Isaacs condition, and so this theorem is a direct consequence of Theorem 8.2.

Thus even when G fails to satisfy the Isaacs condition (36) we may introduce relaxed controls and obtain a value to the game $V_{12} = V_{12}^+ = V_{12}^-$.

THEOREM 8.4 $V^- \leq V_{12} \leq V^+$.

PROOF

We prove $V_{12} \leq V^+$. First we show that $V^+ = V_1^+$; we consider the game H_N^+ (see Theorem 4.3). Letting $\delta = 2^{-N}$ and $t_j = j\delta$ for $0 \leq j \leq 2^N$, we consider the game $H_N^+(t_k,x)$ as the game H_N^+ with initial condition $x(t_k) = x$. The dynamics of $H_N^+(t_k,x)$ are given by (19) and the pay-off by

$$P(y(t),z(t)) = g(x(1)) + \sum_{j=k+1}^{2^N} \int_{t_j}^{t_j} h(t_{j-1}, x(t_{j-1}), y(t), z(t))\,dt. \quad (44)$$

It is easy to see that $H_N^+(t_k,x)$ has a value $S_N^+(t_k,x)$ and that we have the relationship

$$S_N^+(t_k,x) = \inf_{z(t)} \sup_{y(t)} \left\{ \int_{t_k}^{t_{k+1}} h(t_k,x,y(t),z(t))dt + S_N^+(t_{k+1},x') \right\}. \quad (45)$$

where

$$x' = x + \int_{t_k}^{t_{k+1}} f(t_k,x,y(t),z(t))dt\,.$$

Similarly we consider G_1 in place of G. We obtain

$$S_{N,1}^+(t_k,x) = \inf_{z(t)} \sup_{y(t)} \left\{ \int_{t_k}^{t_{k+1}} h(t_k,x,\sigma(t),z(t))dt + S_{N,1}^+(t_{k+1},x') \right\}. \quad (46)$$

where

$$x' = x + \int_{t_k}^{t_{k+1}} f(t_k,x,\sigma(t),z(t))dt.$$

We shall show that for any k, x

$$S_{N,1}^+(t_k,x) = S_N^+(t_k,x)\,;$$

we observe that

$$S_{N,1}^+(1,x) = S_N^+(1,x) = g(x)$$

and proceed by induction. Suppose that for all x

$$S_{N,1}^+(t_{k+1},x) = S_N^+(t_{k+1},x)\,;$$

then for a fixed relaxed control function $\sigma(t)$ on (t_k,t_{k+1}), we may by Corollary 4.5 of [3] determine a sequence $y_n(t)$ of control functions on

(t_k, t_{k+1}) such that

$$\int_{t_k}^{t_{k+1}} f(t_k, x, y_n(t), z(t))\,dt \to \int_{t_k}^{t_{k+1}} h(t_k, x, \sigma(t), z(t))\,dt$$

for any control function $z(t)$ for J_2. We thus obtain from (45) and (46)

$$S^+_{N,1}(t_k, x) = S^+_N(t_k, x).$$

It follows that

$$S^+_{N,1} = S^+_N$$

and so by Theorem 4.3

$$V^+ = V^+_1.$$

Now clearly

$$V_{12} = V^+_{12} \leq V^+_1$$

so that

$$V_{12} \leq V^+.$$

9 THE EXISTENCE OF VALUE WITHOUT CONDITIONS (F1) - (F5)

In this section we dispense with conditions (F1) - (F5) in Theorem 8.2. However, we still require the pay-off in G to be of the form (18); the relaxation of this condition is given in §10. Let us first observe, as by Lemma 3.2 the set of all $x \in R^m$ which are 'attainable' by a trajectory in G is bounded, that we may assume that the functions f, g and h vanish outside compact sets. We shall show first that f and h may each be approximated by functions satisfying conditions (F).

LEMMA 9.1 *If* $f: I \times R^m \times Y \times Z \to R^m$ *is continuous and of compact support, and satisfies a Lipschitz condition in* x,

$$\|f(t,x_1,y,z) - f(t,x_2,y,z)\| \leq k(t)\|x_1 - x_2\|$$

where

$$\int_0^1 k(t) = A < \infty ,$$

then there is a sequence of continuous functions $f^{(n)}: I \times R^m \times Y \times Z \to R^m$ *such that*

(i) $f^{(n)}$ *satisfies a Lipschitz condition in* t *for each* n *,*

$$\|f^{(n)}(t_1,x,y,z) - f^{(n)}(t_2,x,y,z)\| \leq Q^{(n)}|t_1 - t_2| ,$$

(ii) $f^{(n)}$ *satisfies a Lipschitz condition in* x

$$\|f^{(n)}(t,x_1,y,z) - f^{(n)}(t,x_2,y,z)\| \leq k^{(n)}(t)\|x_1 - x_2\| ,$$

where

$$\sup_{0 \leq t \leq 1} |k^{(n)}(t)| = k_n < \infty ,$$

and

$$\int_0^1 k^{(n)}(t)dt \leq 6A \quad \textit{each} \quad n ,$$

(iii) $f^{(n)} \to f$ *uniformly .*

PROOF

Let $\delta = 2^{-n}$ and $t_j = j\delta$ $0 \leq j \leq 2^n$; we choose s_j for $j = 1,2,\dots,2^n$ such that

(a) $t_{j-1} < s_j \leq t_j$

(b) $$\frac{k(s_j)}{2^n} \leq 2\int_{t_{j-1}}^{t_j} k(t)\,dt \tag{47}$$

Then we define for $0 \le t \le s_1$

$$f^{(n)}(t,x,y,z) = f(s_1,x,y,z) ;$$

for $s_j \le t \le s_{j+1}$ $\quad j = 1,2,\dots,2^n - 1$

$$f^{(n)}(t,x,y_1,z) = \frac{s_{j+1} - t}{s_{j+1} - s_j} f(s_j,x,y_1,z) + \frac{t - s_j}{s_{j+1} - s_j} f(s_{j+1},x,y_1,z) ;$$

for $t > s_{2^n}$

$$f^{(n)}(t,x,y,z) = f(s_{2^n},x,y,z) .$$

It is clear, by using the uniform continuity of f (f has compact support) that $f^{(n)} \to f$ uniformly. Secondly $f^{(n)}$ is Lipschitz in t for each n. Finally the functions $k^{(n)}(t)$ in (ii) may be chosen such that

$$k^{(n)}(t) = k(s_1) \qquad 0 \le t \le s_1 ;$$

$$k^{(n)}(t) \le \max\{k(s_j),k(s_{j+1})\} \quad s_j \le t \le s_{j+1} ;$$

$$k^{(n)}(t) = k(s_{2^n}) \qquad s_{2^n} \le t \le 1 ;$$

so that $\int_0^1 k^{(n)}(t)dt \le \sum_{j=1}^{2^n} k(s_j)(s_{j+1} - s_{j-1})$

where $s_0 = 0$ and $s_{2^n+1} = 1$.

$$\int_0^1 k^{(n)}(t)dt \le \sum_{j=1}^{2^n} k(s_j) \frac{3}{2^n}$$

$$\le 6A \qquad \text{by (47)} .$$

THEOREM 9.2 *Let* G *be a differential game satisfying only (1) - (3) and with a pay-off given by (18). Then if* G *satisfies the Isaacs condition (38)*

$$F^+(t,x,p) = F^-(t,x,p)$$

for $(t,x,p) \in I \times R^m \times R^m$, *then* $V^+ = V^-$.

(i.e. G has a value in the sense of Friedman [10].

PROOF

We prove Theorem 9.2 initially under the assumption that g satisfies the conditions (F3) and (F5). We also make the assumption throughout the proof that f, g and h vanish outside some compact set, (this was justified in the opening remarks of this section).

Using Lemma 9.1 we construct a sequence $f^{(n)}$ of functions converging to f uniformly and satisfying (i) - (iii) of 9.1. We may also determine a sequence of functions $h^{(n)}: I \times R^m \times Y \times Z \to R$ converging uniformly to h and each satisfying (F2) and (F4). For convenience we write $f = f^{(\infty)}$ and $h = h^{(\infty)}$.

Then for $1 \leq m \leq \infty$ and $1 \leq n \leq \infty$, we consider the game $G^{(m,n)}$ with dynamics

$$\frac{dx}{dt} = f^{(n)}(t,x,y,z) \tag{48}$$

initial condition

$$x(0) = 0$$

and pay-off

$$P^{(m,n)}(y(t),z(t)) = g(x(1)) + \int_0^1 h^{(m)}(t,x(t),y(t),z(t))dt . \tag{49}$$

(Thus $G \equiv G^{(\infty,\infty)}$.)

We shall use the following notation:

(i) g satisfies (F3) with constant M^*;

(ii) $h^{(m)}$ satisfies (F2) with constant M_m;

(iii) $\sup_{t,x,y,z} \|f(t,x,y,z) - f^{(n)}(t,x,y,z)\| = \eta_n$;

(iv) $\sup_{t,x,y,z} |h(t,x,y,z) - h^{(m)}(t,x,y,z)| = \varepsilon_m$.

Now let $(y(t),z(t))$ be any pair of control functions and let $1 \leq m < \infty$ and $1 \leq n < \infty$; let $x(t)$ be the trajectory determined by $(y(t),z(t))$ in $G^{(m,\infty)}$ and let $x'(t)$ be the trajectory in $G^{(m,n)}$.

Then

$$\left\|\frac{dx'(t)}{dt} - \frac{dx(t)}{dt}\right\| \leq \eta_n + \|f(t,x'(t),y(t),z(t)) - f(t,x(t),y(t),z(t))\|$$

$$\leq \eta_n + k(t)\|x'(t) - x(t)\|$$

and arguing as in Theorem 3.4 or Lemma 3.1 we obtain for $0 \leq t \leq 1$

$$\|x'(t) - x(t)\| \leq \eta_n e^A.$$

Then

$$\int_0^1 |h^{(m)}(t,x'(t),y(t),z(t)) - h^{(m)}(t,x(t),y(t),z(t))|\,dt$$

$$\leq \int_0^1 M_m \eta_n e^A dt$$

$$= M_m \eta_n e^A$$

while

$$|g(x'(1)) - g(x(1))| \leq M^* \eta_n e^A$$

so that

$$|P^{(m,n)}(y(t),z(t)) - P^{(m,\infty)}(y(t),z(t))| \leq (M^* + M_m)\eta_n e^A.$$

Thus if α is a pseudo-strategy for J_1 (see §2), then we compare its values $u^{(m,n)}(\alpha)$ and $u^{(m,\infty)}(\alpha)$ in $G^{(m,n)}$ and $G^{(m,\infty)}$ to obtain

$$|u^{(m,n)}(\alpha) - u^{(m,\infty)}(\alpha)| \leq (M^* + M_m)\eta_n e^A$$

and hence we have

$$|U^{(m,n)}(s) - U^{(m,\infty)}(s)| \leq (M^* + M_m)\eta_n e^A$$

where

$$U^{(m,n)}(s) = \sup_{\alpha \in \Gamma(s)} u^{(m,n)}(\alpha).$$

Hence by Theorem 3.1

$$|V^{+(m,n)} - V^{+(m,\infty)}| \leq (M^* + M_m)\eta_n e^A \tag{50}$$

and

$$|V^{-(m,n)} - V^{-(m,\infty)}| \leq (M^* + M_m)\eta_n e^A \tag{51}$$

where $V^{+(m,n)}$ is the upper Friedman value of $G^{(m,n)}$, etc.

A similar but simpler argument gives us

$$|V^{+(m,\infty)} - V^+| \leq \varepsilon_m \tag{52}$$

$$|V^{-(m,\infty)} - V^-| \leq \varepsilon_m \tag{53}$$

In $G^{(m,n)}$ consider the Isaacs functions

$$F^{+(m,n)}(t,x,p) = \min_{z\in Z}\max_{y\in Y} (p.f^{(n)} + h^{(m)})$$

$$F^{-(m,n)}(t,x,p) = \max_{y\in Y}\min_{z\in Z} (p.f^{(n)} + h^{(m)}) .$$

We have

$$|(p.f^{(n)} + h^{(m)}) - (p.f + h)| \leq \|p\|\eta_n + \varepsilon_m$$

for any $(t,x,p,y,z) \in I \times R^m \times R^m \times Y \times Z$ and so we may deduce

$$|F^{+(m,n)}(t,x,p) - F^+(t,x,p)| \leq \|p\|\eta_n + \varepsilon_m$$

$$|F^{-(m,n)}(t,x,p) - F^-(t,x,p)| \leq \|p\|\eta_n + \varepsilon_m$$

and, as G satisfies the Isaacs condition (36), we have

$$|F^{+(m,n)}(t,x,p) - F^{-(m,n)}(t,x,p)| \leq 2(\|p\|\eta_n + \varepsilon_m) .$$

The game $G^{(m,n)}$ satisfies (F1) - (F5) and so by Lemma 6.2 we have

$$|W^{+(m,n)} - W^{-(m,n)}| \leq 2e(e^{12A}(M_m + M^*)\eta_n + \varepsilon_m)$$

(note that by Lemma 9.1 $G^{(m,n)}$ satisfies (2) and (3) with A replaced by 6A).

Hence by Theorem 8.1

$$|V^{+(m,n)} - V^{-(m,n)}| \le 2e(M_m + M^*)e^{12A}\eta_n + \varepsilon_m).$$

By (50), (51), (52) and (53) we obtain

$$|V^+ - V^-| \le 2e(M_m + M^*)e^{12A}\eta_n + \varepsilon_m) + 2\varepsilon_m + 2(M_m^1 + M^*)\eta_n e^A,$$

for all $1 \le m < \infty$, $1 \le n < \infty$.

Letting $n \to \infty$ we obtain, keeping m fixed

$$|V^+ - V^-| \le 2\varepsilon_m(e+1)$$

and so letting $m \to \infty$

$$V^+ = V^-.$$

It remains only to remove the conditions on the function g. Suppose, then, that g does not satisfy (F3) and (F5); then we may take a sequence $g^{(m)}$ of functions satisfying (F3) and (F5) and such that $g^{(m)} \to g$ uniformly (we recall that g has compact support). Consider the game $G_{(m)}$ with dynamics given by (1) and pay-off

$$P_{(m)}(y(t),z(t)) = g^{(m)}(x(1)) + \int_0^1 h(t,x(t),y(t),z(t))dt. \tag{54}$$

Then as

$$P_{(m)}(y(t),z(t)) \to P(y(t),z(t))$$

uniformly in the control functions $(y(t),z(t))$, by the standard argument, applied above, we obtain

$$V^+_{(m)} \to V^+$$

$$V^-_{(m)} \to V^-.$$

However

$$V^+_{(m)} = V^-_{(m)} \quad \text{for each } m,$$

and so

$$V^+ = V^-.$$

10. GAMES WITH GENERAL PAY-OFF

In this section we refine Theorem 9.2 further by replacing g by a general functional μ on the Banach space of trajectories. For a fixed function f satisfying (1) - (3) we observe that the set X of trajectories is relatively compact in $[C(I)]^m$ (Lemma 3.3). Let us consider the Banach space $C(\overline{X})$ of all continuous real-valued functions on $\overline{X}$ (in the uniform norm); by the Stone-Weierstrass Theorem the set of functions $\rho \in C(\overline{X})$ of the form

$$\rho(x) = g(x(t_1),\ldots,x(t_n)) \tag{55}$$

with $t_n = 1$, is dense in $C(\overline{X})$.

THEOREM 10.1 *Let* G *be the differential game defined by equations (1) - (4). Then if* G *satisfies the Isaacs condition (38), we have* $V^+ = V^-$, *i.e.* G *possesses a value in the sense of Friedman* [10].

PROOF

The functional μ by the preceding remarks may be approximated uniformly *on the set of trajectories* by the functional of the type (55). By the argument used in Theorem 9.2 it is only necessary to establish the result for functionals of this type; the full result will then follow by an approximation procedure. We proceed by induction on the number n in (55); suppose that when $k \leq n$ and

$$\mu(x) = g(x(t_1),\ldots,x(t_k))$$

where $t_k = 1$, we have $V^+ = V^-$. Let us now assume

$$\mu(x) = g(x(t_1),\ldots,x(t_n),x(t_{n+1}))$$

where $t_{n+1} = 1$. For fixed $x_1,x_2\ldots x_n \in R^m$

consider the game $G(x_1,\ldots,x_n)$ with dynamics given by (1), initial condition

$$x(t_n) = x_n,$$

and pay-off

$$P = g(x_1,x_2,\ldots,x_n,\ x(1)) + \int_{t_n}^{1} h(t,x(t),y(t),z(t))dt\,. \qquad (56)$$

As G satisfies the Isaacs condition (28) for $0 \leq t \leq 1$, so does $G(x_1,\ldots,x_n)$ for $t_n \leq t \leq 1$, and hence by Theorem 9.2, $G(x_1,\ldots,x_n)$ has a value $V(x_1,\ldots,x_n) = V^+(x_1,\ldots,x_n) = V^-(x_1,\ldots,x_n)$. We show that $V(x_1,\ldots,x_n)$ is a continuous function on $(R^m)^n$. Suppose that $(x_1,\ldots,x_n) \in (R^m)^n$; then for $\varepsilon > 0$, there exists $\delta > 0$ such that

(i) whenever $\|x - x'\| \leq \delta$, then

$$|h(t,x,y,z) - h(t,x',y,z)| \leq \varepsilon/_2\,,$$

(ii) whenever $\|x_k - x_k'\| \leq \delta$ $k = 1,2,\ldots,n+1$

$$|g(x_1',\ldots,x_{n+1}') - g(x_1,\ldots,x_{n+1})| \leq \varepsilon/_2\,.$$

(We assume as in Theorem 9.2 that h vanishes outside some compact set, so that (i) is justified by uniform continuity of h).

Now suppose

$$\|x_k' - x_k\| \leq \delta e^{-A} \qquad k = 1,2,\ldots,n$$

where

$$A = \int_0^1 k(t)dt\,.$$

Let $(y(t),z(t))$ be any pair of control functions on $[t_n,1]$ determining trajectories $x(t)$ in $G(x_1,\ldots,x_n)$ and $x'(t)$ in $G(x_1',\ldots,x_n')$. Then we have

$$\|x'(t) - x(t)\| \leq \delta e^{-A} \exp\left\{\int_{t_n}^{t} k(s)\,ds\right\}$$

$$\leq \delta \qquad t_n \leq t \leq 1 .$$

Hence

$$\left|\int_{t_n}^{1} h(t,x(t),y(t),z(t)) - h(t,x'(t),y(t),z(t))dt\right| \leq \varepsilon/2$$

and

$$|g(x_1,\ldots,x_n,x(1)) - g(x_1',\ldots,x_n',x'(1))| \leq \varepsilon/2 .$$

Therefore the pay-offs $P(y,z)$ in $G(x_1,\ldots,x_n)$ and $P'(y,z)$ in $G(x_1',\ldots,x_n')$ satisfy

$$|P - P'| \leq \varepsilon .$$

A standard argument on pseudo-strategies as in Theorem 7.1 yields

$$|V(x_1,\ldots,x_n) - V(x_1',\ldots,x_n')| \leq \varepsilon ,$$

so that V is continuous.

We may treat the function $U(s)$ for $G(x_1,\ldots,x_n)$ in a similar manner; we denote this by $U(s,x,\ldots,x_n)$, and obtain U is continuous in $x_1,\ldots,x_n$. We also have as $s \downarrow 0$

$$\lim_{s\to 0} U(s,x_1,\ldots,x_n) = V(x_1,\ldots,x_n)$$

monotonically. By Dini's Theorem

$$\lim_{s\to 0} U(s,x_1,\ldots,x_n) = V(x_1,\ldots,x_n) \tag{57}$$

uniformly on compacta.

Let us now consider the game $G_0^{(s)}$ with dynamics given by

$$\frac{dx}{dt} = f(t,x(t),y(t),z(t))$$

initial condition

$$x(0) = 0$$

and pay-off

$$P = U(s,x(t_1),\ldots,x(t_n)) + \int_0^{t_n} h(t,x(t),y(t),z(t))\,dt. \tag{58}$$

By the inductive hypothesis $G_0^{(s)}$ has a value which we denote by $\tilde{U}(s)$, and by the usual argument, equation (57) yields that

$$\lim_{s\to 0} \tilde{U}(s) = V_0, \tag{59}$$

where V_0 is the value of the game G_0 whose dynamics and initial condition are as in $G_0^{(s)}$, but whose pay-off is given by

$$P = V(x(t_1)\ \ldots\ x(t_n)) + \int_0^{t_n} h(t,x(t),y(t),z(t))\,dt.$$

We now consider the sets M_1 and M_2 of control functions (modulo functions equal almost everywhere) as direct sums of control functions on $[0,t_n]$ and $[t_n,1]$.

Thus if $z(t) \in M_2$, then we may write $z = [z^{(1)},z^{(2)}]$ where $z^{(1)}: [0,t_n] \to Z$. We write N_{21} for the space of measurable functions $z^{(1)}: [0,t_n] \to Z$ and N_{22} for the space of measurable functions $z^{(2)}: [t_n,1] \to Z$ (so that, in the sense above $M_2 = N_{21} \oplus N_{22}$). Similarly $M_1 = N_{11} \oplus N_{12}$.

For $s > 0$, an s-delay strategy α for J_1 in the game G is a map

$$\alpha: M_2 \to M_1$$

satisfying (6), and induces a map

$$\bar{\alpha}: N_{21} \to N_{11},$$

where $\bar{\alpha}$ is an s-delay strategy in $G_0^{(s)}$.

For fixed $z(t) \in N_{21}$ we define $\alpha_z : N_{22} \to N_{12}$ by

$$\alpha[z,w] = [\bar{\alpha}z, \alpha_2 w] .$$

Then $(\bar{\alpha}z, z)$ determine a trajectory $x(t)$, $0 \leq t \leq t_n$, and α_z is an s-delay strategy in $G(x(t_1),\dots,x(t_n))$.

$$\inf_{w \in N_{22}} P\left[\alpha(z,w),(z,w)\right] = \int_0^{t_n} h(t,x(t),\bar{\alpha}z(t),z(t))\,dt$$

$$+ \inf_{w \in N_{22}} \left\{ \int_{t_n}^{1} h(t,x(t),a_z w(t),w(t))\,dt + g(x(t_1)\dots x(t_n),x(1)) \right\}$$

$$= u(\alpha_z) + \int_0^{t_n} h(t,x(t),\bar{\alpha}z(t),z(t))\,dt$$

$$\leq U(s,x(t_1)\dots x(t_n)) + \int_0^{t_n} h(t,x(t),\bar{\alpha}z(t)z(t))\,dt$$

so that

$$u(\alpha) \leq u_s(\bar{\alpha})$$

where $u_s(\alpha)$ is the value of $\bar{\alpha}$ in $G_0^{(s)}$. Thus we have

$$U(s) \leq \tilde{U}(s)$$

and letting $s \to 0$, by (49)

$$V^- \leq V_0 .$$

Conversely, suppose $x_1 \dots x_n$ are given; then there exists an s-delay strategy for J_1, $\alpha^*(x_1 \dots x_n)$, for $s < 1 - t_n$, in $G(x_1 \dots x_n)$ such that

$$u(\alpha^*(x_1 \dots x_n)) \geq U(s,x_1 \dots x_n) - s .$$

Given an s-delay strategy α in G we define $\hat{\alpha}$ by

$$\hat{\alpha}[z,w] = [\bar{\alpha}z, \hat{\alpha}_z w]$$

where

$$\begin{aligned}\hat{\alpha}_z w &= \alpha([z,w])(t) \qquad t_n \le t \le t_n + s \\ &= \alpha^*(x(t_1)\dots x(t_n))w(t) \qquad t_n + s \le t \le 1,\end{aligned}$$

and $x(t)$, $0 \le t \le t_n$, is the trajectory determined by

$$(\bar{\alpha}z, z).$$

Suppose $\hat{x}(t)$ and $x^*(t)$, $t_n \le t \le 1$, are the trajectories corresponding to $(\hat{\alpha}_z w, w)$ and $(\alpha^*(x(t_1)\dots x(t_n))w, w)$. Then we may show, by a method similar to that in Theorem 3.4 that

$$\|\hat{x}(t) - x^*(t)\| \le 2B'' se^A$$

where

$$B'' = \sup\|f(t,x,y,z)\|$$

(as in Theorem 9.2, we can assume that f is bounded, since the set of attainable $x \in R^m$ is relatively compact). Hence we have

$$|u(\hat{\alpha}_z) - u(\alpha^*(x(t_1)\dots x(t_n)))| \le \eta(s)$$

where $\eta(s) \to 0$ as $s \to 0$. Then

$$\begin{aligned}\inf_{w \in N_{21}} P\left[\hat{\alpha}[z,w],\ [z,w]\right] &= u(\hat{\alpha}_z) + \int_0^{t_n} h(t,x(t),\bar{\alpha}z(t),z(t))dt \\ &\ge U(s,x(t_1)\dots x(t_n)) - \eta(s) - s \\ &\quad + \int_0^{t_n} h(t,x(t),\bar{\alpha}z(t),z(t))dt\end{aligned}$$

and so

$$u(\hat{\alpha}) \ge u_s(\bar{\alpha}) - s - \eta(s).$$

Hence as $\hat{\alpha}$ is an s-delay strategy we may deduce

$$U(s) \geq \tilde{U}(s) - s - \eta(s)$$

so that, taking limits as $s \to 0$

$$V^{-} \geq \lim_{s \to 0} (\tilde{U}(s) - s - \eta(s))$$
$$= V_0$$

by (59). Hence $V^{-} = V_0$ and we similarly conclude that $V^{+} = V_0$.

11. OPTIMAL STRATEGIES AND SADDLE POINTS

The problem which naturally follows the proof of the existence of value for differential games satisfying Isaacs condition is that of determining the existence of optimal strategies for the two players. In the setting of relaxed controls we may show that each player possesses a strategy achieving the value of the game (in relaxed controls).

We treat therefore the game G_{12}, which by Theorem 10.1 possesses a value, as the Isaacs condition is always satisfied in relaxed controls. We consider the space M_1^* and M_2^* of relaxed control functions for the two players (modulo functions equal almost everywhere); following [3] we assign topologies to M_1^* and M_2^*, so that M_1^* is identified as a subset of the dual of the Banach space $L^1(C(Y))$ of integrable functions $\varphi: I \to C(Y)$ in the weak* topology. The duality is given by

$$\langle \varphi(t), \sigma(t) \rangle = \int_0^1 \left\{ \int_Y \varphi(t)\, d\sigma(t) \right\} dt \tag{60}$$

where $\sigma(t)$ is a relaxed control function. It follows that M_1^* is compact in this topology (and also that M_1 is dense in M_1^*). Similarly M_2^* is compact as a subset of $[L^1(C(Z))]^*$.

LEMMA 11.1 *The pay-off function* $P: M_1^* \times M_2^* \to R$ *is separately continuous.*

PROOF

See [3] (an example to show that P is not necessarily jointly continuous is also given).

THEOREM 11.2 *In* G_{12}, J_1 *possesses a strategy* α *and* J_2 *a strategy* β *such that*

$$u(\alpha) = u(\beta) = V_{12}.$$

PROOF

We prove 11.2 only for J_1. Let α_n be a strategy for J_1 such that

$$u(\alpha_n) \geq V_{12} - \frac{1}{n}.$$

We identify α_n as a member of $(M_1^*)^{M_2^*}$ with the product topology; by Tychonoff's theorem this space is compact and so we may produce a subset (α_λ) of (α_n) such that α_λ converges to some α in $(M_1^*)^{M_2^*}$. For $\tau(t) \in M_2^*$

$$P(\alpha_\lambda \tau, \tau) \geq u(\alpha_\lambda)$$

and so by Lemma 11.1

$$P(\alpha\tau, \tau) \geq \limsup_{\lambda \to \infty} u(\alpha_\lambda)$$

$$\geq V_{12}.$$

Then

$$u(\alpha) \geq V_{12}$$

and hence

$$u(\alpha) = V_{12}.$$

We also can see that $\alpha: M_2^* \to M_1^*$ is indeed a strategy, for if

$$\tau_1(t) = \tau_2(t) \qquad \text{a.e.} \quad 0 \le t \le T$$

then

$$\alpha_\lambda \tau_1(t) = \alpha_\lambda \tau_2(t) \qquad \text{a.e.} \quad 0 \le t \le T$$

Suppose $\varphi: [0,T] \to C(Y)$ is integrable; then we have

$$\lim_\lambda \langle \bar{\varphi}, \alpha_\lambda \tau_1 \rangle = \lim_\lambda \langle \bar{\varphi}, \alpha_\lambda \tau_2 \rangle$$

where

$$\bar{\varphi}(t) = \varphi(t)$$

for $0 \le t \le T$, and zero for $T \le t \le 1$.

Hence

$$\langle \bar{\varphi}, \alpha\tau_1 \rangle = \langle \bar{\varphi}, \alpha\tau_2 \rangle$$

and as this is true for all such φ we have, taking

$$\varphi(t,y) = \theta(t)\psi(y)$$

for $\theta \in L^1(I)$ and $\psi \in C(Y)$,

$$\int_0^T \theta(t) \int_Y \psi(y) d\alpha\tau_1(t,y) = \int_0^T \theta(t) \int_Y \psi(y) d\alpha\tau_2(t,y) .$$

So

$$\int_Y \psi(y)\, d\alpha\tau_1(t,y) = \int_Y \psi(y) d\alpha\tau_2(t,y) \qquad \text{a.e.} \quad 0 \le t \le T.$$

As in Lemma 2.1, using the metrizability of Y, we may deduce from this, that

$$\alpha\tau_1(t) = \alpha\tau_2(t) \qquad \text{a.e.} \quad 0 \le t \le T.$$

THEOREM 11.3 *In* G_1, J_1 *possesses an optimal strategy* α *such that* $u(\alpha) = V_1^+$.

PROOF Identical.

Although we have produced optimal strategies in G_{12}, it does not follow that we have any reasonable saddle point solution. For, although, α and β have the same value, we cannot necessarily (see Example 2.2) produce an outcome to the game if J_1 elects to use α and J_2 to use β. There is only one case in which we have any guarantee of an outcome to two strategies. If α and β are assumed to be continuous for the given topologies on M_1^* and M_2^*, then $\beta\alpha: M_2^* \to M_2^*$ is continuous and maps a compact convex subset of $L^1(C(Z))^*$ into itself; we may therefore quote the Schauder fixed point theorem to deduce the existence of $\tau(t)$ such that

$$\beta\alpha\{\tau(t)\} = \tau(t).$$

Then $(\alpha\tau(t),\tau(t))$ is the outcome of the strategies α and β. However, it is unlikely that the optimal strategies of 11.2 are continuous.

To circumvent these difficulties we introduce a weaker saddle point idea. We shall say that a sequence (α_n) of $\frac{1}{n}$- delay strategies (i.e. $\alpha_n \in \Gamma(\frac{1}{n})$) for J_1 in G is an *approximate strategy;* similarly a sequence (β_n) where $\beta_n \in \Delta(\frac{1}{n})$ is an approximate strategy for J_2. It is easy to verify that there exists a unique pair of control functions $(y_n(t),z_n(t))$ such that

$$\alpha_n z_n(t) = y_n(t)$$

$$\beta_n y_n(t) = z_n(t).$$

This is done by an induction process - first α_n determines the control function $y_n(t)$ for $0 \leq t \leq \frac{1}{n}$, which in turn determines $z_n(t)$ for $0 \leq t \leq \frac{2}{n}$, etc. We define the *pay-off* $P\{(\alpha_n),(\beta_n)\}$ as the set of all limit points of the sequence $P\{y_n(t),z_n(t)\}$. We shall say that the sequences (α_n^*) and (β_n^*) from a *saddle point for approximate strategies* if for any approximate strategies (α_n), (β_n)

$$P\{(\alpha_n),(\beta_n^*)\} \leq P\{(\alpha_n^*),(\beta_n^*)\} \leq P\{(\alpha_n^*),(\beta_n)\}$$

(where for sets of real numbers $\mathcal{A}$ and $\mathcal{B}$ we have $\mathcal{A} \leq \mathcal{B}$ if and only if $a \leq b$ for $a \in \mathcal{A}$, $b \in \mathcal{B}$).

THEOREM 11.4 *If* G *satisfies the Isaacs condition then there exists a saddle point for approximate strategies in* G .

PROOF

By Theorem 10.1 G has a value V and we may determine $\alpha_n^* \in \Gamma(\frac{1}{n})$ and $\beta_n^* \in \Delta(\frac{1}{n})$ with

$$u(\alpha_n^*) = V - \varepsilon_n$$

$$v(\beta_n^*) = V + \delta_n$$

where $\varepsilon_n \to 0$ and $\delta_n \to 0$. Then if

$$\alpha_n^* z_n^* = y_n^*$$

$$\beta_n^* y_n^* = z_n^*$$

$$u(\alpha_n^*) \leq P(y_n^*, z_n^*) \leq v(\beta_n^*) ,$$

so that

$$\lim_{n\to\infty} P(y_n^*, z_n^*) = V ,$$

i.e.

$$P\{(\alpha_n^*),(\beta_n^*)\} = \{V\} .$$

For any other approximate strategy (α_n) for J_1, suppose

$$\alpha_n z_n = y_n$$

$$\beta_n^* y_n = z_n$$

so that

$$P(y_n, z_n) \leq v(\beta_n)$$
$$= V + \delta_n$$

and so

$$P\{(\alpha_n), (\beta_n^*)\} \leq \{V\}.$$

The result follows easily.

Clearly the method of 11.4 shows that a saddle point over the appropriate strategies exists if and only if the game G has a value in the sense of Friedman; see Friedman [10] for alternative notions of saddle point.

12. RELATIONSHIP WITH THE WORK OF FRIEDMAN

In this section we relate our results to those of Friedman [10]; he makes the assumption both f and h split in y and z, i.e.

$$f(t,x,y,z) = f^1(t,x,y) + f^2(t,x,z) \tag{61}$$

and

$$h(t,x,y,z) = h^1(t,x,y) + h^2(t,x,z). \tag{62}$$

Under assumptions (61) and (62), the Isaacs condition (38) holds automatically, and so Theorem 10.1 shows that the game G will have a value. Thus our Theorem 10.1 includes the main theorem of [10].

However the approach used by Friedman is much simpler than the one adopted here. The main improvement due to the splitting of (61) and (62) is expressed in the following lemma.

LEMMA 12.1 *Let* α *be a pseudo-strategy for* J_1 *and for* $0 < \delta \leq 1$ *define*

$$\alpha_\delta z(t) = \alpha z(t - \delta) \qquad t \geq \delta$$

$$= y_o(t) \qquad t < \delta$$

where $y_o(t)$ *is some fixed control function. Then*

$$\lim_{\delta \to 0} P(\alpha_\delta z(t), z(t)) = P(\alpha z(t), z(t))$$

uniformly in α *and* $z(t)$ *when (61) and (62) are satisfied.*

We omit the proof, but remark that intuitively this is a clear consequence of (61) and (62). The slight reaction time δ is unimportant since the effects of y and z do not interact; contrast the situation in Example 2.2 where the delay of an optimal strategy may be disastrous.

THEOREM 12.2 *Assuming* (61) *and* (62), *the functions* $U(s)$ *and* $V(s)$ *are continuous. In particular*

$$U(s) = V(-s) \quad \textit{for any} \quad -1 \leq s \leq 1 .$$

PROOF

From Lemma 12.1 we conclude that

$$u(\alpha_\delta) \to u(\alpha)$$

for any pseudo-strategy α. If $\alpha \in \Gamma(s)$ then $\alpha_\delta \in \Gamma(s+\delta)$, and so

$$u(\alpha_\delta) \leq U(s+\delta) \leq U(s)$$

and

$$\lim_{\delta \to 0} \sup_{\alpha} u(\alpha_\delta) = U(s)$$

so that

$$\lim_{\delta \to 0} U(s+\delta) = U(s)$$

and so U is continuous on the right. To show that U is continuous on the left we need an anticipatory version of Lemma 12.2; define

$$\alpha_{-\delta} z(t) = \alpha z(t+\delta) \qquad t \leq 1-\delta$$

$$= y_o(t) \qquad t > 1-\delta .$$

Then as in Lemma 12.1

$$\lim_{\delta \to 0} P(\alpha_{-\delta} z(t), z(t)) = P(\alpha z(t), z(t))$$

uniformly in $z(t)$ and α. It follows immediately that U is continuous.

The implication $U(s) = V(-s)$ is deduced as in the remarks at the end of §3.

Another simplification due to (61) and (62) worth observing is (see Lemma 11.1):

PROPOSITION 12.3 *Assuming* (61) *and* (62) *the pay-off in* G_{12} $P: M_1^* \times M_2^* \to R$ *is jointly continuous.*

We omit the details of the proof of this result. The main step is to prove that if $\sigma_n(t) \to \sigma(t)$ in M_1^* and $\tau_n(t) \to \tau(t)$ in M_2^* then the corresponding trajectories $x_n(t)$ converge uniformly to the trajectory $x(t)$ induced by $\sigma(t)$ and $\tau(t)$; this is very similar to the result proved by Warga [20].

The assumptions (61) and (62) do allow us under certain circumstances to show the existence of an optimal strategy for J_1. We assume also that the set

$$K(t,x) = \left\{ \begin{pmatrix} f^1(t,x,y) \\ h^1(t,x,y) \end{pmatrix} ; \quad y \in Y \right\}$$

is convex in R^{m+1} for each $(t,x) \in I \times R^m$, and also that Y is a subset of some Euclidean space R^q.

THEOREM 12.4 *Under the assumptions above, there is an optimal strategy* α *for* J_1 *such that* $u(\alpha) = V(=V^+ = V^-)$.

PROOF

By Theorem 11.3 there is an optimal strategy α^* for J_1 in G_1; we shall construct a strategy α in G with the same effect using Zorn's Lemma. Suppose that $\alpha: M_2 \to M_1$ has been determined for a subset M of M_2 and that $z(=z(t)) \in M_2 - M$; let

$$T = \sup\{s; \exists z' \in M, z'(t) = z(t) \text{ a.e. } 0 \leq t \leq s\}.$$

Then we may determine $z_n' \in M$ such that

$$z(t) = z_n'(t) \text{ a.e. } 0 \leq t \leq (1-\frac{1}{n})T,$$

and let

$$\alpha z(t) = \alpha z_n'(t) \text{ a.e. } (1-\frac{1}{n-1})T \leq t \leq (1-\frac{1}{n})T \quad n = 2,3,\ldots,$$

This will determine $\alpha z(t)$ for $0 \leq t < T$.

Now $(\alpha^* z,z)$ determines a trajectory $x(t)$ in G_1 and for $t \geq T$

$$\begin{pmatrix} f^1(t,x(t),\alpha^* z(t)) \\ h^1(t,x(t),\alpha^* z(t)) \end{pmatrix} \in K(t,x(t)).$$

By the Filippov Implicit Function Theorem, we may determine a measurable function $y(t)$, $t \geq T$, such that

$$f^1(t_1\, x(t),\, y(t)) = f^1(t_1\, x(t),\alpha^* z(t))$$
$$h^1(t_1\, x(t),\, y(t)) = h^1(t_1\, x(t),\alpha^* z(t)).$$

Then we define

$$\alpha z(t) = y(t) \qquad t \geq T.$$

It is clear that $(\alpha z,z)$ determines the same trajectory and pay-off as $(\alpha^* z,z)$. Using Zorn's Lemma we build up α in this way and clearly α is a strategy with

$$u(\alpha) = u(\alpha^*) = V.$$

BIBLIOGRAPHY

1. BERKOVITZ, L.D. 'A variational approach to differential games'. In 'Advances in Game Theory', Annals of Math. Study 52, pp 127-174, Princeton University Press, Princeton 1964.

2. ——— 'A differential game with no pure strategy solution'. loc. cit. pp 175-194.

3. ELLIOTT, R.J., KALTON, N.J. and MARKUS, L. 'Saddle points for linear differential games', J. S.I.A.M., Series A, Control, to appear.

4. FLEMING, W.H. 'The convergence problem for differential games'. J. Math. Analysis and Applications 3 (1961) pp 102-116.

5. ——— 'The convergence problem for differential games. II' In 'Advances in Game Theory', Annals of Math. Study 52 pp 195-210. Princeton University Press, Princeton 1964.

6. ——— 'The Cauchy problem for degenerate parabolic equations', J. Math. and Mechanics 13 (1964) pp 987-1008.

7. ——— 'Non-linear partial differential equations - probabilistic and game theoretic methods' Centro Internazionale Matematico Estivo IV Ciclo Varenna, Agosto 1970 Coordinatore G. Prodi. Edizioni Cremonese, Roma 1971.

8. FRIEDMAN, A. 'On quasi-linear parabolic equations of the second order. II' J. Math. and Mechanics 9 (1960) pp 539-556.

9. ——— 'Partial differential equations of parabolic type' Prentice Hall, Englewood Cliffs N.J. 1964.

10. ——— 'On the definition of differential games and the existence of value and saddle points'. J. Diff. Eqns. 7(1970), pp 69-91.

11. ——— 'Existence of Value and saddle points for differential games of pursuit and evasion'. J. Diff. Eqns. 7(1970) pp 92-110.

12. ——— 'Existence of value and saddle points for differential games of survival'. J. Diff. Eqns 7(1970) pp 111-125.

13. ISAACS, R. 'Differential Games'. John Wiley and Sons, New York, London, 1965.

14. McSHANE, E.J. 'Relaxed controls and variational problems'. J. S.I.A.M. Ser. A., Control, 5(1967), pp 438-485.

15. OLEINIK, O.A. and KRUZHKOV, S.N. 'Quasilinear parabolic equations of second order with many independent variables', Uspekhi Mat. Nauk, 16(1961) pp 115-155.

16. ROXIN, E. 'The Axiomatic approach in differential games'. J. Optimization Theory and Applications. 3(1969) pp 153-163.

17. VARAIYA, P. and LIN, J. 'Existence of saddle points in differential games'. J. S.I.A.M. Series A, Control, 7(1969) pp 141-157.

18. von NEUMANN. J. and MORGENSTERN, O. 'Theory of Games and Economic Behaviour'. Princeton University Press, Princeton, 2nd ed. 1947.

19. WALD, A. 'Statistical Decision Functions'. John Wiley and Sons, New York, London. 1950.

20. WARGA, J. 'Functions of relaxed controls'. J. S.I.A.M. Series A, Control, 5(1967), pp 628-641.

21. YOUNG, L.C. 'Generalized curves and the existence of an attained absolute minimum in the calculus of variations', C.R. Soc. Sci. Lettres Varsovie, CL111, 30(1937) pp 212-234.

Mathematics Institute
University of Warwick.

MEMO/126